Trigonometry

Trigonometry

Casey Swanson

Larsen & Keller
www.larsen-keller.com

Trigonometry
Casey Swanson
ISBN: 978-1-64172-135-6 (Hardback)

Larsen & Keller
Published by Larsen and Keller Education,
5 Penn Plaza,
19th Floor,
New York, NY 10001, USA

Cataloging-in-Publication Data

Trigonometry / edited by Casey Swanson.
p. cm.
Includes bibliographical references and index.
ISBN 978-1-64172-135-6
1. Trigonometry. 2. Geometry. I. Swanson, Casey.
QA531 .T75 2019
516.24--dc23

Printed and bound in China.

For more information regarding Larsen and Keller Education and its products, please visit the publisher's website www.larsen-keller.com

Table of Contents

Preface

Trigonometry is an important field of mathematics. It deals with the study of relationship between angles and lengths of triangles. Such studies have led to the development of trigonometric functions. A trigonometric function relates the angles of a triangle to the lengths of the sides. Sine, cosine and tangents are the most prominent trigonometric functions. Such functions are useful in the study of triangles and for modeling periodic phenomena. The study of planar right-angle triangles is generally dealt in trigonometry. These triangles are two-dimensional triangles, which have one angle equal to 90 degrees. Spherical trigonometry studies the triangles on spheres. Trigonometry and trigonometric functions are applied to a wide array of fields such as optics, astronomy, medical imaging, cartography, economies, seismology, etc. This book provides comprehensive insights into the field of trigonometry. Such selected concepts that redefine this field have been presented herein. Coherent flow of topics, student-friendly language and extensive use of examples make this book an invaluable source of knowledge.

A detailed account of the significant topics covered in this book is provided below:

Chapter 1, The branch of mathematics that deals with the relationships between the lengths and the angles of triangles is referred to as trigonometry. This chapter will discuss briefly all the significant aspects of trigonometry such as right triangle trigonometry, trigonometric integral and trigonometric series, etc. **Chapter 2**, Trigonometric functions are the functions of an angle which relates the angles of a triangle with the lengths of its sides. This chapter discusses in detail the different trigonometric functions namely the sine function, cosine function, tangent function, cosine function, secant function and cotangent function. **Chapter 3**, Trigonometric functions are periodic, which means that they repeat their values in regular periods. The understanding of trigonometric functions is incomplete without a study of the plotting of their graphs on a XY plane. This chapter provides a detailed discussion of the graphing of periodic trigonometric functions. **Chapter 4**, Trigonometry is an important area of mathematics that is governed by a number of laws and theorems. This chapter closely examines the fundamental laws and theorems in trigonometry to provide an extensive understanding of the subject. It includes topics like law of sines, law of cosines, law of tangents, Pythagorean theorem, Thales's theorem, etc. **Chapter 5**, Trigonometric identities are based on trigonometric functions. They are equalities that hold for every value of a variable. The topics elaborated in this chapter will help in gaining a better perspective about the different trigonometric identities such as tangent half-angle formula, Lagrange's trigonometric identities, Euler's formula, De Moivre's formula, Morrie's law, etc. **Chapter 6**, An equation which involves one or more trigonometric ratios of unknown angles is known as a trigonometric equation. The aim of this chapter is to explore trigonometric equations and the methods for finding the solutions of such equations. **Chapter 7**, Trigonometry has applications in the fields of astronomy, architecture, engineering, geodesy, physical sciences, seismology, etc. The diverse uses and applications of trigonometry in different areas have been thoroughly discussed in this chapter. It also includes topics like application of right triangle, application of radian measure, application of trigonometric graphs, etc.

I would like to make a special mention of my publisher who considered me worthy of this opportunity and also supported me throughout the process. I would also like to thank the editing team at the back-end who extended their help whenever required.

Casey Swanson

1

Introduction to Trigonometry

The branch of mathematics that deals with the relationships between the lengths and the angles of triangles is referred to as trigonometry. This chapter will discuss briefly all the significant aspects of trigonometry such as right triangle trigonometry, trigonometric integral and trigonometric series, etc.

Trigonometry is defined as one of the branches of mathematics that deals with the relationships that involve lengths and also the angles of triangles. In a simple manner, we can say that Trigonometry is the study of triangles. The word trigonometry was derived from the Greek word, where, 'TRI' means Three, 'GON' means sides and whereas the 'METRON' ways to measure. This concept is given to us by a Greek mathematician Hipparchus.

To be more specific, trigonometry is all about a right-angled triangle, where one of the internal angles measure about 90°. Moreover, it is one of those divisions in mathematics that helps in finding the angles and missing sides of a triangle. In trigonometry, the angles are either measured in radians or degrees.

This branch divides into two sub-branches called plane trigonometry and the spherical geometry. Trigonometry, in general, is about the trigonometric formulas, trigonometric ratios, and functions, Right-Angled Triangles, etc.

Plane Trigonometry

In many applications of trigonometry the essential problem is the solution of triangles. If enough sides and angles are known, the remaining sides and angles as well as the area can be calculated, and the triangle is then said to be solved. Triangles can be solved by the law of sines and the law of cosines. To secure symmetry in the writing of these laws, the angles of the triangle are lettered A, B, and C and the lengths of the sides opposite the angles are lettered a, b, and c, respectively.

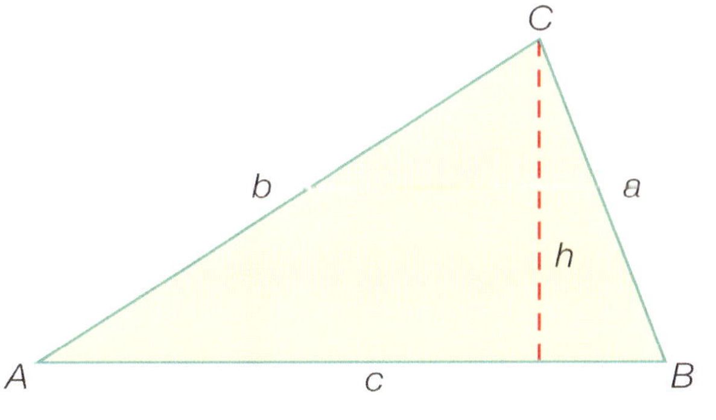

Standard lettering of a triangleIn addition to the angles (A, B, C) and sides (a, b, c), one of the three heights of the triangle (h) is included by drawing the line segment from one of the triangle's vertices (in this case C) that is perpendicular to the opposite side of the triangle..

The law of sines is expressed as an equality involving three sine functions while the law of cosines is an identification of the cosine with an algebraic expression formed from the lengths of sides opposite the corresponding angles. To solve a triangle, all the known values are substituted into equations expressing the laws of sines and cosines, and the equations are solved for the unknown quantities. For example, the law of sines is employed when two angles and a side are known or when two sides and an angle opposite one are known. Similarly, the law of cosines is appropriate when two sides and an included angle are known or three sides are known.

Spherical Trigonometry

Spherical trigonometry involves the study of spherical triangles, which are formed by the intersection of three great circle arcs on the surface of a sphere. Spherical triangles were subject to intense study from antiquity because of their usefulness in navigation, cartography, and astronomy.

The angles of a spherical triangle are defined by the angle of intersection of the corresponding tangent lines to each vertex. The sum of the angles of a spherical triangle is always greater than the sum of the angles in a planar triangle (π radians, equivalent to two right angles). The amount by which each spherical triangle exceeds two right angles (in radians) is known as its spherical excess. The area of a spherical triangle is given by the product of its spherical excess E and the square of the radius r of the sphere it resides on—in symbols, Er^2.

Law of sines: $\dfrac{\sin a}{\sin A}=\dfrac{\sin b}{\sin B}=\dfrac{\sin c}{\sin C}$

Law of cosines:
$$\cos a=\cos b\cos c+\sin b\sin c\cos A$$
$$\cos b=\cos a\cos c+\sin a\sin c\cos B$$
$$\cos c=\cos a\cos b+\sin a\sin b\cos C$$

Half-angle formulas:
$$\tan\left(\frac{A}{2}\right)=\sqrt{\frac{\sin(s-b)\sin(s-c)}{\sin s\sin(s-a)}}$$
$$\tan\left(\frac{B}{2}\right)=\sqrt{\frac{\sin(s-c)\sin(s-a)}{\sin s\sin(s-b)}}$$
$$\tan\left(\frac{C}{2}\right)=\sqrt{\frac{\sin(s-a)\sin(s-b)}{\sin s\sin(s-c)}},\text{where } s=\frac{a+b+c}{2}$$

Half-side formulas:
$$\tan\left(\frac{a}{2}\right)=\sqrt{\frac{-\cos S\cos(S-A)}{\cos(S-B)\cos(S-C)}}$$
$$\tan\left(\frac{b}{2}\right)=\sqrt{\frac{-\cos S\cos(S-B)}{\cos(S-A)\cos(S-C)}}$$
$$\tan\left(\frac{c}{2}\right)=\sqrt{\frac{-\cos S\cos(S-C)}{\cos(S-A)\cos(S-B)}},\text{where } S=\frac{A+B+C}{2}$$

By connecting the vertices of a spherical triangle with the centre O of the sphere that it resides on, a special "angle" known as a trihedral angle is formed. The central angles (also known as dihedral angles) between each pair of line segments OA, OB, and OC are labeled α, β, and γ to correspond to the sides (arcs) of the spherical triangle labeled a, b, and c, respectively. Because a trigonometric function of a central angle and its corresponding arc have the same value, spherical trigonometry formulas are given in terms of the spherical angles A, B, and C and, interchangeably, in terms of the arcs a, b, and c and the dihedral angles α, β, and γ. Furthermore, most formulas from plane trigonometry have an analogous representation in spherical trigonometry. For example, there is a spherical law of sines and a spherical law of cosines.

As was described for a plane triangle, the known values involving a spherical triangle are substituted in the analogous spherical trigonometry formulas, such as the laws of sines and cosines, and the resulting equations are then solved for the unknown quantities.

$$tan\left(\frac{a}{2}\right)cos\left(\frac{B-C}{2}\right)=tan\left(\frac{b+c}{2}\right)cos\left(\frac{B+C}{2}\right)$$

$$tan\left(\frac{a}{2}\right)sin\left(\frac{B-C}{2}\right)=tan\left(\frac{b-c}{2}\right)sin\left(\frac{B+C}{2}\right)$$

$$cot\left(\frac{A}{2}\right)cos\left(\frac{b-C}{2}\right)=tan\left(\frac{B+C}{2}\right)cos\left(\frac{b+c}{2}\right)$$

$$cot\left(\frac{A}{2}\right)sin\left(\frac{b-C}{2}\right)=tan\left(\frac{B-C}{2}\right)sin\left(\frac{b+c}{2}\right)$$

For each of the included formulas two other formulas can be obtained by permuting the sides and angles: $(a,A)\to(b,B)\to(c,C)\to(a,A)$

Many other relations exist between the sides and angles of a spherical triangle. Worth mentioning are Napier's analogies (derivable from the spherical trigonometry half-angle or half-side formulas), which are particularly well suited for use with logarithmic tables.

Analytic Trigonometry

Analytic trigonometry combines the use of a coordinate system, such as the Cartesian coordinate system used in analytic geometry, with algebraic manipulation of the various trigonometry functions to obtain formulas useful for scientific and engineering applications.

Trigonometric functions of a real variable x are defined by means of the trigonometric functions of an angle. For example, sin x in which x is a real number is defined to have the value of the sine of the angle containing x radians. Similar definitions are made for the other five trigonometric functions of the real variable x. These functions satisfy the previously noted trigonometric relations with A, B, 90°, and 360° replaced by x, y, π/2 radians, and 2π radians, respectively. The minimum period of tan x and cot x is π, and of the other four functions it is 2π.

In calculus it is shown that sin x and cos x are sums of power series. These series may be used to compute the sine and cosine of any angle. For example, to compute the sine of 10°, it is necessary to find the value of sin $\pi/18$ because 10° is the angle containing $\pi/18$ radians. When $\pi/18$ is substituted in the series for sin x, it is found that the first two terms give 0.17365, which is correct to five decimal places for the sine of 10°. By taking enough terms of the series, any number of decimal places can be correctly obtained. Tables of the functions may be used to sketch the graphs of the functions.

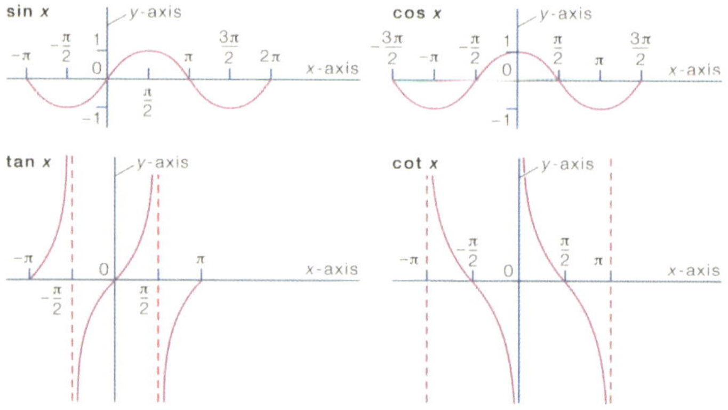

Graphs of some trigonometric functions Note that each of these functions is periodic. Thus, the sine and cosine functions repeat every 2π, and the tangent and cotangent functions repeat every π.

Each trigonometric function has an inverse function, that is, a function that "undoes" the original function. For example, the inverse function for the sine function is written arcsin or $\sin^{-1}$, thus $\sin^{-1}(\sin x) = \sin(\sin^{-1} x) = x$. The other trigonometric inverse functions are defined similarly.

Angle Measurement and Tables

If there is anything that distinguishes trigonometry from the rest of geometry, it is that trig depends on angle measurement and quantities determined by the measure of an angle. Of course, all of geometry depends on treating angles as quantities, but in the rest of geometry, angles aren't measured, they're just compared or added or subtracted.

Trigonometric functions such as sine, cosine, and tangent are used in computations in trigonometry. These functions relate measurements of angles to measurements of associated straight lines.

Trig functions are not easy to compute like polynomials are. So much time goes into computing them in ancient times that tables were made for their values. Even with tables, using trig functions takes time because any use of a trig function involves at least one multiplication or division, and, when several digits are involved, even multiplication and division are slow.

These ratios are given by the following trigonometric functions of the known angle A, where a, b and c refer to the lengths of the sides in the accompanying figure:

- Sine function (sin), defined as the ratio of the side opposite the angle to the hypotenuse.

$$\sin A = \frac{\text{opposite}}{\text{hypotenuse}} = \frac{a}{c}.$$

- Cosine function (cos), defined as the ratio of the adjacent leg to the hypotenuse.

$$\cos A = \frac{\text{adjacent}}{\text{hypotenuse}} = \frac{b}{c}.$$

- Tangent function (tan), defined as the ratio of the opposite leg to the adjacent leg.

$$\tan A = \frac{\text{opposite}}{\text{adjacent}} = \frac{a}{b} = \frac{a}{c} * \frac{c}{b} = \frac{a}{c} / \frac{b}{c} = \frac{\sin A}{\cos A}.$$

The hypotenuse is the side opposite to the 90 degree angle in a right triangle; it is the longest side of the triangle and one of the two sides adjacent to angle A. The adjacent leg is the other side that is adjacent to angle A. The opposite side is the side that is opposite to angle A. The terms perpendicular and base are sometimes used for the opposite and adjacent sides respectively.

The reciprocals of these functions are named the cosecant (csc), secant (sec), and cotangent (cot), respectively:

$$\csc A = \frac{1}{\sin A} = \frac{\text{hypotenuse}}{\text{opposite}} = \frac{c}{a},$$

$$\sec A = \frac{1}{\cos A} = \frac{\text{hypotenuse}}{\text{adjacent}} = \frac{c}{b},$$

$$\cot A = \frac{1}{\tan A} = \frac{\text{adjacent}}{\text{opposite}} = \frac{\cos A}{\sin A} = \frac{b}{a}.$$

The inverse functions are called the arcsine, arccosine, and arctangent, respectively. There are arithmetic relations between these functions, which are known as trigonometric identities. The cosine, cotangent, and cosecant are so named because they are respectively the sine, tangent, and secant of the complementary angle abbreviated to "co-".

With these functions, one can answer virtually all questions about arbitrary triangles by using the law of sines and the law of cosines. These laws can be used to compute the remaining angles and sides of any triangle as soon as two sides and their included angle or two angles and a side or three sides are known. These laws are useful in all branches of geometry, since every polygon may be described as a finite combination of triangles.

Extending the Definitions

The above definitions only apply to angles between 0 and 90 degrees (0 and $\pi/2$ radians). Using the unit circle, one can extend them to all positive and negative arguments. The trigonometric functions are periodic, with a period of 360 degrees or 2π radians. That means their values repeat at those intervals. The tangent and cotangent functions also have a shorter period, of 180 degrees or π radians.

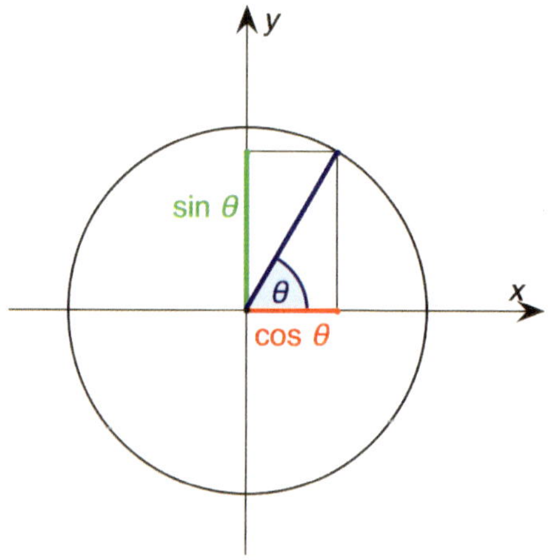

Figure – Sine and cosine of an angle θ defined using the unit circle.

The trigonometric functions can be defined in other ways besides the geometrical definitions above, using tools from calculus and infinite series. With these definitions the trigonometric functions can be defined for complex numbers. The complex exponential function is particularly useful.

$$e^{x+iy} = e^{x}(\cos y + i\sin y)$$

See Euler's and De Moivre's formulas.

	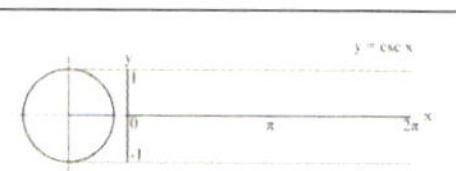	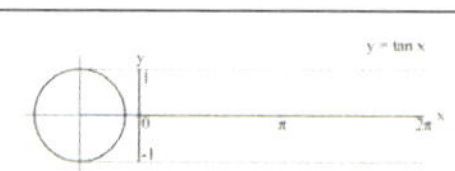
Graphing process of $y = \sin(x)$ using a unit circle.	Graphing process of $y = \csc(x)$, the reciprocal of sine, using a unit circle.	Graphing process of $y = \tan(x)$ using a unit circle.

Mnemonics

A common use of mnemonics is to remember facts and relationships in trigonometry. For example, the *sine*, *cosine*, and *tangent* ratios in a right triangle can be remembered by representing them and their corresponding sides as strings of letters. For instance, a mnemonic is SOH-CAH-TOA:

Sine = Opposite ÷ Hypotenuse

Cosine = Adjacent ÷ Hypotenuse

Tangent = Opposite ÷ Adjacent

One way to remember the letters is to sound them out phonetically. Another method is to expand the letters into a sentence, such as "Some Old Hippie Caught Another Hippie Trippin' On Acid".

Calculating Trigonometric Functions

Trigonometric functions were among the earliest uses for mathematical tables. Such tables were incorporated into mathematics textbooks and students were taught to look up values and how to interpolate between the values listed to get higher accuracy. Slide rules had special scales for trigonometric functions.

Today, scientific calculators have buttons for calculating the main trigonometric functions (sin, cos, tan, and sometimes cis and their inverses). Most allow a choice of angle measurement methods: degrees, radians, and sometimes gradians. Most computer programming languages provide function libraries that include the trigonometric functions. The floating point unit hardware incorporated into the microprocessor chips used in most personal computers has built-in instructions for calculating trigonometric functions.

Applications

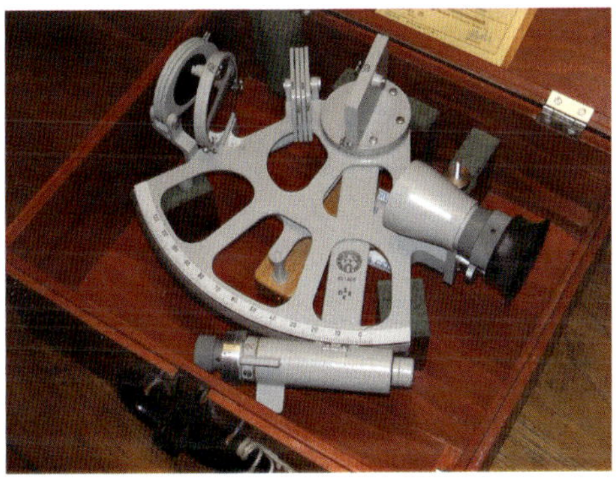

Sextants are used to measure the angle of the sun or stars with respect to the horizon. Using trigonometry and a marine chronometer, the position of the ship can be determined from such measurements.

There is an enormous number of uses of trigonometry and trigonometric functions. For instance, the technique of triangulation is used in astronomy to measure the distance to nearby stars, in geography to measure distances between landmarks, and in satellite navigation systems. The sine and cosine functions are fundamental to the theory of periodic functions, such as those that describe sound and light waves.

Fields that use trigonometry or trigonometric functions include astronomy (especially for locating apparent positions of celestial objects, in which spherical trigonometry is essential) and hence navigation (on the oceans, in aircraft, and in space), music theory, audio synthesis, acoustics, optics, electronics, biology, medical imaging (CT scans and ultrasound), pharmacy, chemistry, number theory (and hence cryptology), seismology, meteorology, oceanography, many physical sciences, land surveying and geodesy, architecture, image compression, phonetics, economics, electrical engineering, mechanical engineering, civil engineering, computer graphics, cartography, crystallography and game development.

Pythagorean Identities

The following identities are related to the Pythagorean theorem and hold for any value:

$$\sin^2 A + \cos^2 A = 1$$

$$\tan^2 A + 1 = \sec^2 A$$

$$\cot^2 A + 1 = \csc^2 A$$

Angle Transformation Formulae

$$\sin(A \pm B) = \sin A \cos B \pm \cos A \sin B$$

$$\cos(A \pm B) = \cos A \cos B \mp \sin A \sin B$$

$$\tan(A \pm B) = \frac{\tan A \pm \tan B}{1 \mp \tan A \tan B}$$

$$\cot(A \pm B) = \frac{\cot A \cot B \mp 1}{\cot B \pm \cot A}$$

Common Formulae

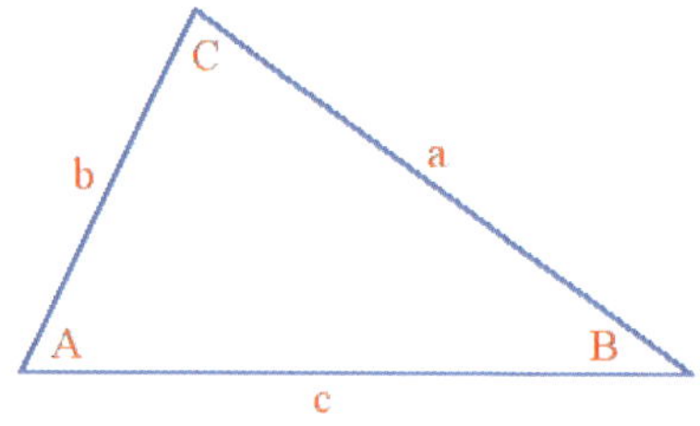

Triangle with sides *a*,*b*,*c* and respectively opposite angles *A*,*B*,*C*

Certain equations involving trigonometric functions are true for all angles and are known as *trigonometric identities*. Some identities equate an expression to a different expression involving the same angles. These are listed in List of trigonometric identities. Triangle identities that relate the sides and angles of a given triangle are listed below.

In the following identities, A, B and C are the angles of a triangle and a, b and c are the lengths of sides of the triangle opposite the respective angles (as shown in the diagram).

Law of Sines

The law of sines (also known as the "sine rule") for an arbitrary triangle states:

$$\frac{a}{\sin A} = \frac{b}{\sin B} = \frac{c}{\sin C} = 2 \quad = \frac{abc}{2},$$

where Δ is the area of the triangle and R is the radius of the circumscribed circle of the triangle:

$$R = \frac{abc}{\sqrt{(a+b+c)(a-b+c)(a+b-c)(b+c-a)}}.$$

Another law involving sines can be used to calculate the area of a triangle. Given two sides a and b and the angle between the sides C, the area of the triangle is given by half the product of the lengths of two sides and the sine of the angle between the two sides:

$$\text{Area} = \Delta = \frac{1}{2} ab \sin C.$$

Law of Cosines

The law of cosines (known as the cosine formula, or the "cos rule") is an extension of the Pythagorean theorem to arbitrary triangles:

$$c^2 = a^2 + b^2 - 2ab\cos C,$$

or equivalently:

$$\cos C = \frac{a^2 + b^2 - c^2}{2ab}.$$

The law of cosines may be used to prove Heron's formula, which is another method that may be used to calculate the area of a triangle. This formula states that if a triangle has sides of lengths a, b, and c, and if the semi perimeter is

$$s = \frac{1}{2}(a+b+c),$$

then the area of the triangle is:

$$\text{Area} = \Delta = \sqrt{s(s-a)(s-b)(s-c)} = \frac{abc}{4R},$$

where R is the radius of the circumcircle of the triangle.

Law of Tangents

The law of tangents:

$$\frac{a-b}{a+b} = \frac{\tan\left[\frac{1}{2}(A-B)\right]}{\tan\left[\frac{1}{2}(A+B)\right]}$$

Euler's Formula

Euler's formula, which states that $e^{ix} = \cos x + i\sin x$, produces the following analytical identities for sine, cosine, and tangent in terms of e and the imaginary unit i:

$$\sin x = \frac{e^{ix} - e^{-ix}}{2i}, \qquad \cos x = \frac{e^{ix} + e^{-ix}}{2}, \qquad \tan x = \frac{i(e^{-ix} - e^{ix})}{e^{ix} + e^{-ix}}.$$

Using Triangles to Measure Circles

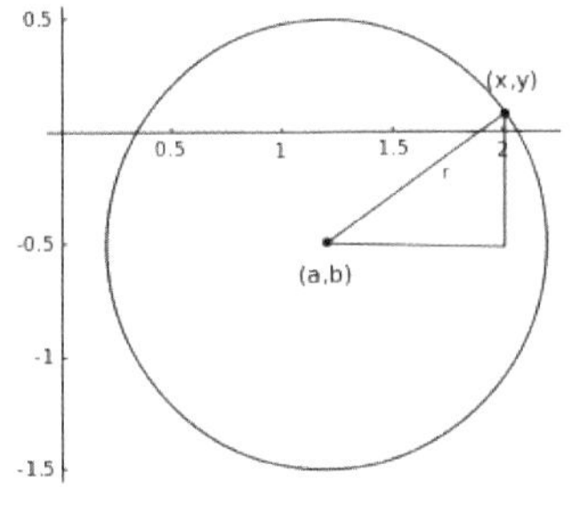

Using a right triangle to define a circle.

Geometric Curves: Conics in Trig

Trigonometry is powerful enough to make measurements of things that are not triangles. Conics such as hyperbolae and ellipses are examples of how awesomely sneaky trigonometry can be - a triangle (and all its formulae) can be hidden inside an oval.

Let's start with a circle. One of the first things one learns in trigonometry is that the radii and arcs of a circle can be found using a right triangle. This is because the hypotenuse of a right triangle is also the slope of the line connecting the center of the circle with a point on the circle (as shown below.) This same point can also be found using the trigonometric functions.

Working with triangles to find information about a circle is easy enough, but what happens with ellipses? They're just flattened circles, but the distance from the center to the edge is not uniform as it is in a circle.

It could be argued that an ellipse is better defined by its foci than its center (while noting that the center is still useful in calculating the equation for the ellipse.) The distance from one focus (F1) to any point (P) added to the distance from the other focus (F2) to point P does not differ as one travels around the ellipse. An ellipse is related using b2 = a2 – c2 where c is the distance from the center to either focus (either positive or negative), a is the distance from the center to the vertex (major axis), and b is the distance from the center to the minor-axis.

$$\frac{(x-h)^2}{a^2}+\frac{(y-k)^2}{b^2}=1$$

Equations for Ellipses

The equation for an ellipse with center (h, k) where the x-axis is the major axis (as in the ellipse shown below) is:

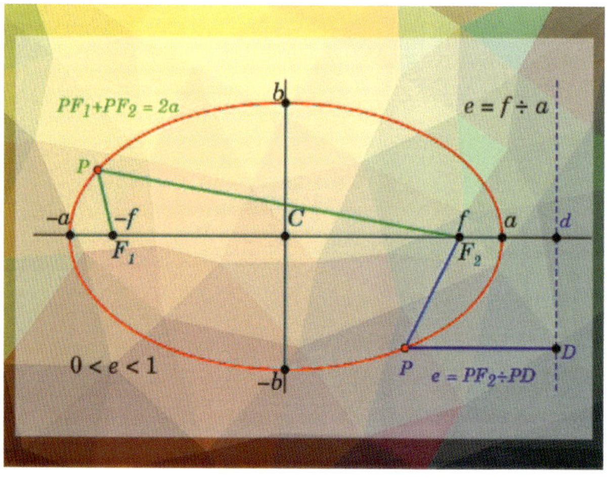

An ellipse where the x-axis is the major axis. Vertices at (h,a) and (h,-a).

$$\frac{(y-k)^2}{a^2}+\frac{(x-h)^2}{b^2}=1$$

However, the equation for an ellipse where the major axis is the y-axis is related by:

Equations for Hyperbolae

A hyperbola looks very different from an ellipse. In fact, nearly oppositely so, it's a hyperbola split in half with the halves facing in opposite directions. However, in terms of finding the equations of hyberbolae versus any other "shape," the two are closely related.

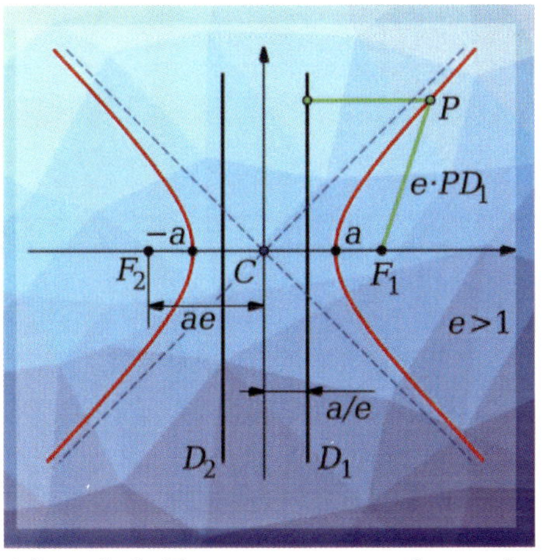

A hyperbola Trans verse across the x-axis.

$$\frac{(x-h)^2}{a^2}-\frac{(y-k)^2}{b^2}=1$$

For x-axis Trans versed hyperbolae

$$\frac{(y-k)^2}{a^2}-\frac{(x-h)^2}{b^2}=1$$

For y-axis Trans versed hyperbolae

Like an ellipse, the center of a hyperbola is referenced by (h,k.) However, a hyperbola only has one vertex (noted by the distance a from the center in either the x or y-direction depending on the transverse axis.)

Also unlike an ellipse, the foci of a hyperbola (noted by distance c from the center) are further from the center than the vertex. The Pythagorean Theorem rears its head here too, where c2 = b2 + a2 using the equations to the right.

As you can see, trigonometry can bring one further than just finding the missing length of a triangle (or a missing angle.) It's used for more than just measuring the height of a tree by the shadow it casts or finding the distance between two buildings given some unusual scenario. Trigonometry can be applied further to define and describe circles and circle-like shapes.

Hyperbolae and ellipses serve as great examples of how trigonometry can quickly deviate from just stating the Pythagorean Theorem and the few relationships between the lengths of the sides of a simple triangle (the trig functions).

Right Triangle Trigonometry

A right angle has a value of 90 degrees (90°). A right triangle is a triangle in which one angle is a right angle. The relation between the sides and angles of a right triangle is the basis for trigonometry.

The side opposite the right angle is called the hypotenuse (side c in the figure). The sides adjacent to the right angle are called legs (sides a and b). Side a may be identified as the side adjacent to angle B and opposed to (or opposite) angle A. Side b is the side adjacent to angle A and opposed to angle B.

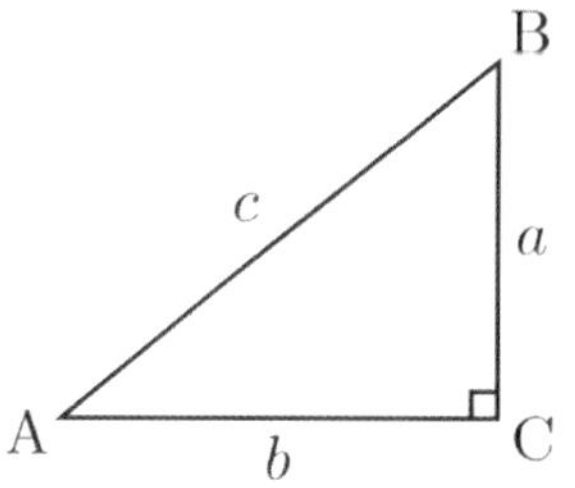

Right triangle: The Pythagorean Theorem can be used to find the value of a missing side length in a right triangle.

If the lengths of all three sides of a right triangle are whole numbers, the triangle is said to be a Pythagorean triangle and its side lengths are collectively known as a Pythagorean triple.

The Pythagorean Theorem

The Pythagorean Theorem, also known as Pythagoras' Theorem, is a fundamental relation in Euclidean geometry. It defines the relationship among the three sides of a right triangle. It states that the square of the hypotenuse (the side opposite the right angle) is equal to the sum of the squares of the other two sides. The theorem can be written as an equation relating the lengths of the sides a, b and c, often called the "Pythagorean equation":

$$a^2 + b^2 = c^2$$

In this equation, c represents the length of the hypotenuse and a and b the lengths of the triangle's other two sides.

Although it is often said that the knowledge of the theorem predates him, the theorem is named after the ancient Greek mathematician Pythagoras (c. 570 – c. 495 BC). He is credited with its first recorded proof.

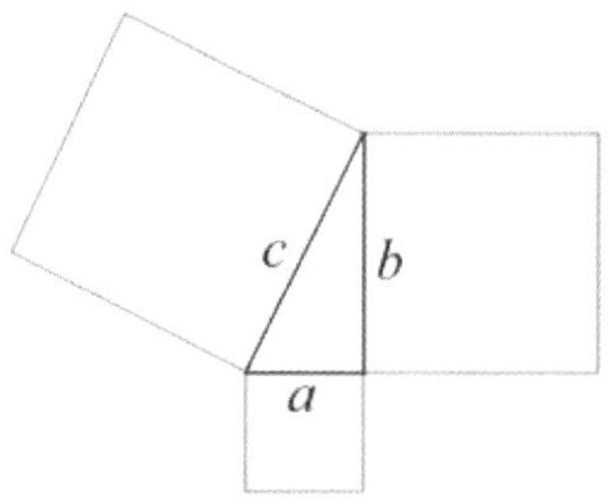

The Pythagorean Theorem: The sum of the areas of the two squares on the legs (a and b) is equal to the area of the square on the hypotenuse (c). The formula is $a^2 + b^2 = c^2$.

Example: A right triangle has a side length of 10feet, and a hypotenuse length of 20 feet. Find the other side length. (round to the nearest tenth of a foot).

Substitute a=10 and c=20 into the Pythagorean Theorem and solve for b.

$$a^2+b^2=c^2$$
$$(10)^2+b^2=(20)^2$$
$$100+b^2=400$$
$$b^2=300$$
$$\sqrt{b^2}=\sqrt{300}$$
$$b=17.3 \text{ feet}$$

Example: A right triangle has side lengths 3 cm and 4 cm. Find the length of the hypotenuse.

Substitute a=3 and b=4 into the Pythagorean Theorem and solve for c.

$$a^2+b^2=c^2$$
$$3^2+4^2=c^2$$
$$9+16=c^2$$
$$25=c^2$$
$$c^2=25$$
$$\sqrt{c^2}=\sqrt{25}$$
$$c=5 \text{ cm}$$

Working of Trigonometric function

Trigonometric functions can be used to solve for missing side lengths in right triangles.

Trigonometric Functions

We can define the trigonometric functions in terms an angle t and the lengths of the sides of the triangle. The adjacent side is the side closest to the angle. (Adjacent means "next to.") The opposite side is the side across from the angle. The hypotenuse is the side of the triangle opposite the right angle, and it is the longest.

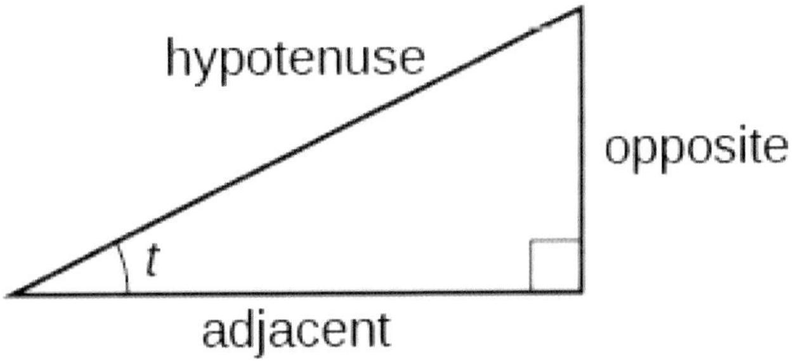

Right triangle: The sides of a right triangle in relation to angle t.

When solving for a missing side of a right triangle, but the only given information is an acute angle measurement and a side length, use the trigonometric functions listed below:

- sine $\sin t = \dfrac{opposite}{hypotenuse}$
- Cosine $cos\, t = \dfrac{adjacent}{hypotenuse}$
- Tangent $tan\, t = \dfrac{opposite}{adjacent}$

The trigonometric functions are equal to ratios that relate certain side lengths of a right triangle. When solving for a missing side, the first step is to identify what sides and what angle are given, and then select the appropriate function to use to solve the problem.

Evaluating a Trigonometric Function of a Right Triangle

Sometimes you know the length of one side of a triangle and an angle, and need to find other measurements. Use one of the trigonometric functions (sin, cos, tan), identify the sides and angle given, set up the equation and use the calculator and algebra to find the missing side length.

Example:

Given a right triangle with acute angle of 34° and a hypotenuse length of 25 feet, find the length of the side opposite the acute angle (round to the nearest tenth):

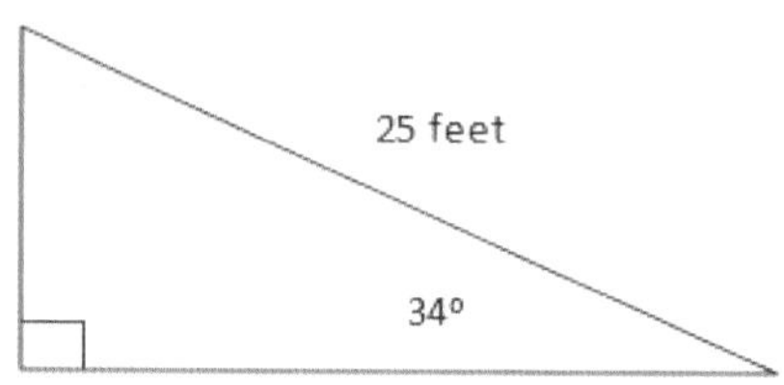

Right triangle: Given a right triangle with acute angle of 34 degrees and a hypotenuse length of 25 feet, find the opposite side length.

Looking at the figure, solve for the side opposite the acute angle of 34 degrees. The ratio of the sides would be the opposite side and the hypotenuse. The ratio that relates those two sides is the sine function.

$$\sin t = \frac{opposite}{hypotenuse}$$

$$\sin(34^\circ) = \frac{x}{25}$$

$$25 . \sin(34^\circ) = x$$

$$x = 25 . \sin(34^\circ)$$

$$x = 25.(0.559...)$$

$$x = 14.0$$

The side opposite the acute angle is 14.014.0 feet.

Example:

Given a right triangle with an acute angle of 83° and a hypotenuse length of 300300 feet, find the hypotenuse length (round to the nearest tenth):

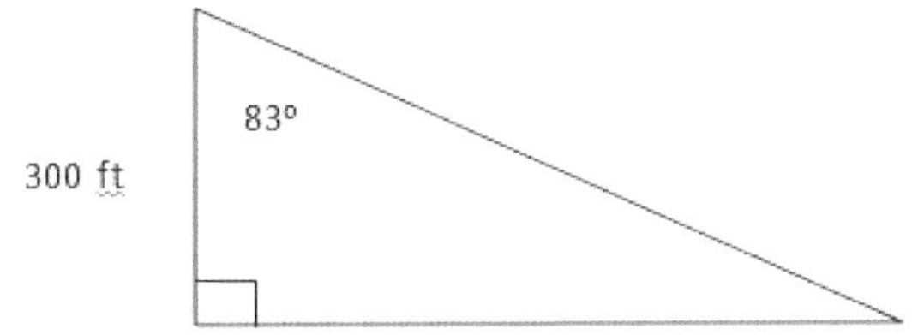

Right Triangle: Given a right triangle with an acute angle of 8383 degrees and a hypotenuse length of 300 feet, find the hypotenuse length.

Looking at the figure, solve for the hypotenuse to the acute angle of 83 degrees. The ratio of the sides would be the *adjacent* side and the *hypotenuse.* The ratio that relates these two sides is the cosine function.

$$\cos t = \frac{adjacent}{hypotenuse}$$

$$cos(83^\circ) = \frac{300}{x}$$

$$x \cdot cos(83^\circ) = 300$$

$$x = \frac{300}{cos(83^\circ)}$$

$$x = \frac{300}{(0.1218\ldots)}$$

$$x = 2461.7\ feet$$

Trigonometric Integral

A trigonometric integral is an integral involving products and powers of trigonometric functions: cosine, sine, tangent, secant, cosecant, and cotangent. Many of these integrals can be handled with u-substitution.

Integrals of the form $\int \sin^m x . \cos^n x\, dx$

Case 1: m is an odd integer:

Step 1: Write $\sin^m x$ as $\sin^{m-1} x . \sin x$.

Step 2: Apply identity: $\sin^2 x = 1 - \cos^2 x$

Step 3: Use the substitution $u = \cos x$.

Example: Evaluate the following integral

$$\int \sin^3 x \cdot \cos^2 x \, dx$$

Solution:

In this example m = 3 and n = 2. Because m is an odd integer we have:

Step 1:

$$\int \sin^5 x \cdot \cos^2 x \, dx = \int \sin^4 x \cdot \sin x \cdot \cos^2 x dx$$

Step 2: Apply identity: $\sin^2 x = 1 - \cos^2 x$. We continue

$$\int \sin^4 x \cdot \sin x \cdot \cos^2 x \, dx = \int \left(\sin^2 x\right)^2 \cdot \sin x \cdot \cos^2 x \, dx = \int \left(1 - \cos^2 x\right)^2 \cdot \cos^2 x \cdot \sin x \, dx$$

Step 3: Use the substitution $u = \cos x$.

If $u = \cos x$, then $du = -\sin x dx$. Now, we have:

$$\int \left(1-\cos^2 x\right)^2 \cdot \cos^2 x \cdot \sin x \, dx \frac{u = \cos x}{du = -\sin x dx} \int \left(1-u^2\right)^2 \cdot u^2 \, du =$$

$$= \int \left(1 - 2u^2 + u^4\right) \cdot u^2 \, du = \int u^2 - 2u^4 + u^6 du =$$

$$= \frac{u^3}{3} - 2\frac{u^5}{5} + \frac{u^7}{7} + C = \frac{1}{3}(\cos x)^3 - \frac{2}{5}(\cos x)^5 + \frac{1}{7}(\cos x)^7 + C$$

Case 2: n is an odd integer:

Step 1: Write $\cos^m x$ as $\cos^{m-1} x \,.\, \cos x$.

Step 2: Apply identity: $\cos^2 x = 1 - \sin^2 x$

Step 3: Use the substitution $u = \sin x$.

Example: Evaluate the following integral

$$\int \sin^6 x. \cos^7 x \, dx$$

Solution:

In this example m = 6 and n = 7. Because n is an odd integer we have:

Step 1:

$$\int \sin^6 x \cdot \cos^7 x \, dx = \int \sin^6 x \cdot \cos^6 x \cdot \cos x \, dx$$

Step 2: Apply identity: $\cos^2 x = 1 - \sin^2 x$.

$$\int sin^6 x \cdot cos^6 x \cdot cos\, x \, dx = \int sin^6 x \cdot \left(cos^2 x\right)^3 \cdot cos\, x \, dx =$$

$$= \int sin^6 x \cdot \left(1 - sin^2\right)^3 \cdot cos\, x \, dx$$

Step 3: Use the substitution $u \quad \sin x$.

If $u = \sin x$, then $du = \cos x . dx$. Now, we have:

$$\int \sin^6 x\left(1-\sin^2 x\right)^3 \cdot \cos x\, dx \overset{u=\sin x}{\underset{du=\cos x\,dx}{=\!=\!=\!=\!=\!=}} \int u^6\left(1-u^2\right)^3 .du =$$

$$= \int u^6 \cdot \left(1-3u^2+3u^4-u^6\right) du = \int u^6 - 3u^8 + 3u^{10} - u^{12} du =$$

$$= \frac{u^7}{7} - 3\frac{u^9}{9} + 3\frac{u^{11}}{11} - \frac{u^{13}}{13} + C =$$

$$= \frac{1}{7}(\sin x)^7 - \frac{1}{3}(\sin x)^9 + \frac{3}{11}(\sin x)^{11} - \frac{1}{13}(\sin x)^{13} + C$$

Case 3: If both m and n are even, we use identities:

$$\sin^2 x = \frac{1-\cos 2x}{2} \qquad \cos^2 x \frac{1+\cos 2x}{2} \qquad \sin x \cos x = \frac{\sin 2x}{2}$$

Example: Evaluate the following integral

$$\int \sin^2 x \cdot \cos^2 x\, dx$$

Solution:

In this example m = 2 and n = 2. Because both m and n are even, we use above identities.

$$\int \sin^2 x . \cos^2 x\, dx = \int \frac{1-\cos 2x}{2} . \frac{1+\cos 2x}{2} dx =$$

$$= \frac{1}{4}\int (1-\cos 2x)(1+\cos 2x)\, dx \overset{2x=u}{\underset{dx=\frac{1}{2}du}{=\!=\!=\!=\!=}} = \frac{1}{8}\int (1-\cos u).(1+\cos u) du$$

$$= \frac{1}{8}\int 1-\cos^2 u\, du = \frac{1}{8}\int 1 - \frac{1+\cos 2u}{2} du = \frac{1}{16}\int 1 - \cos 2u\, du \overset{2u=t}{\underset{du=\frac{1}{2}dt}{=\!=\!=\!=\!=\!=}} \frac{1}{32}\int 2 - \cos t dt =$$

$$= \frac{1}{32}(2t - \sin t) + C = \frac{1}{32}(2.2u - \sin 2u) + C = \frac{1}{32}(4.2x - \sin 4x) + C = \frac{1}{32}(8x - \sin 4x) + C$$

Sine Integral

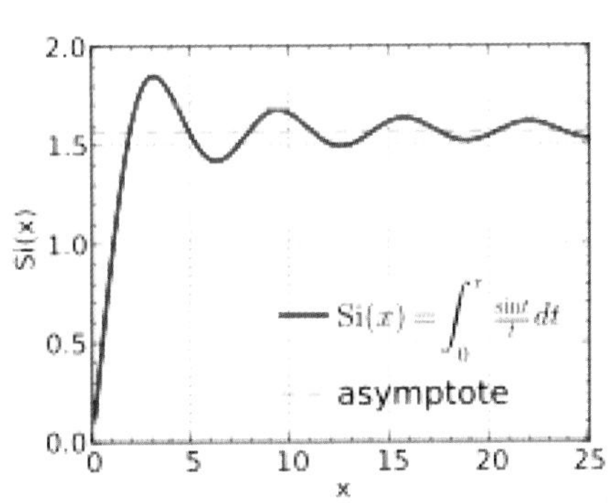

Plot of Si(x) for $0 \le x \le 8\,\pi$.

The different sine integral definitions are

$$\mathrm{Si}(x) = \int_0^x \frac{\sin t}{t} dt$$

$$\mathrm{si}(x) = -\int_x^\infty \frac{\sin t}{t} dt.$$

Note that the integrand sin x / x is the sinc function, and also the zeroth spherical Bessel function. Since sinc is an even entire function (holomorphic over the entire complex plane), Si is entire, odd, and the integral in its definition can be taken along any path connecting the endpoints.

By definition, Si(x) is the antiderivative of sin x / x which is zero for $x = 0$, and si(x) is the antiderivative of sin x / x which is zero for $x = \infty$. Their difference is given by the Dirichlet integral,

$$\mathrm{Si}(x) - \mathrm{si}(x) = \int_0^\infty \frac{\sin t}{t} dt = \frac{\pi}{2}.$$

In signal processing, the oscillations of the sine integral cause overshoot and ringing artifacts when using the sinc filter, and frequency domain ringing if using a truncated sinc filter as a low-pass filter.

Related is the Gibbs phenomenon: if the sine integral is considered as the convolution of the sinc function with the heaviside step function, this corresponds to truncating the Fourier series, which is the cause of the Gibbs phenomenon.

Cosine Integral

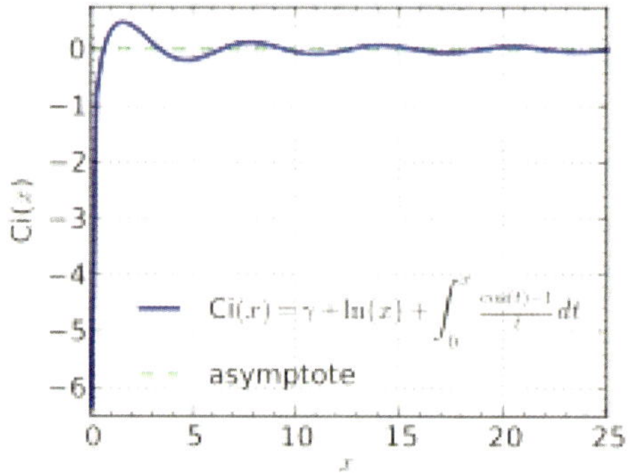

Plot of Ci(x) for $0 < x \leq 8\pi$.

The different cosine integral definitions are

$$\mathrm{Ci}(x) = -\int_x^\infty \frac{\cos t}{t} dt = \gamma + \ln x + \int_0^x \frac{\cos t - 1}{t} \quad (|\operatorname{Arg}(x)| < \pi)$$

$$\mathrm{Cin}(x) = \int_0^x \frac{1 - \cos t}{t} dt,$$

where γ is the Euler–Mascheroni constant. Some texts use ci instead of Ci.

Ci(x) is the antiderivative of $\cos x / x$ (which vanishes at $x \to \infty$). The two definitions are related by

$$\mathrm{Cin}(x) = \gamma + \ln x - \mathrm{Ci}(x).$$

Cin is an even entire function.

Hyperbolic Sine Integral

The hyperbolic sine integral is defined as

$$\mathrm{Shi}(x) = \int_0^x \frac{\sinh(t)}{t} dt.$$

It is related to the ordinary sine integral by

$$\mathrm{Si}(ix) = i\,\mathrm{Shi}(x).$$

Hyperbolic Cosine Integral

The hyperbolic cosine integral is

$$\mathrm{Chi}(x) = \gamma + \ln x + \int_0^x \frac{\cosh t - 1}{t} dt \qquad (|\operatorname{Arg}(x)| < \pi)$$

where γ is the Euler–Mascheroni constant.

It has the series expansion

$$\mathrm{Chi}(x) = \gamma + \ln(x) + \frac{1}{4}x^2 + \frac{1}{96}x^4 + \frac{1}{4320}x^6 + \frac{1}{322560}x^8 + \frac{1}{36288000}x^{10} + O(x^{12}).$$

Auxiliary functions

Trigonometric integrals can be understood in terms of the so-called "auxiliary functions"

$$f(x) \equiv \int_0^\infty \frac{\sin(t)}{t+x} dt = \int_0^\infty \frac{e^{-xt}}{t^2+1} dt = \mathrm{Ci}(x)\sin(x) + \left[\frac{\pi}{2} - \mathrm{Si}(x)\right]\cos(x)$$

$$g(x) \equiv \int_0^\infty \frac{\cos(t)}{t+x} dt = \int_0^\infty \frac{te^{-xt}}{t^2+1} dt = -\mathrm{Ci}(x)\cos(x) + \left[\frac{\pi}{2} - \mathrm{Si}(x)\right]\sin(x).$$

Using these functions, the trigonometric integrals may be re-expressed.

$$\begin{aligned}\mathrm{Si}(x) &= \frac{\pi}{2} - f(x)\cos(x) - g(x)\sin(x) \\ \mathrm{Ci}(x) &= f(x)\sin(x) - g(x)\cos(x).\end{aligned}$$

Nielsen's Spiral

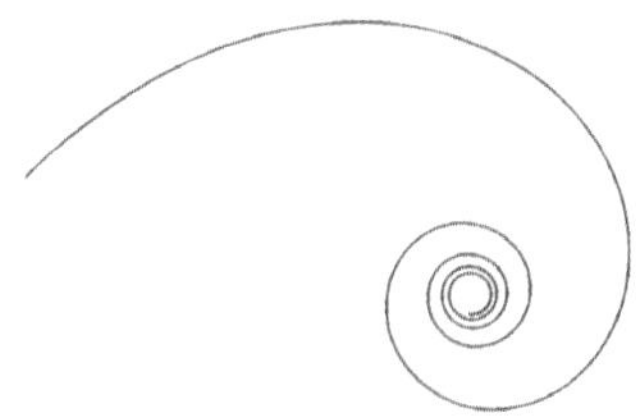

Nielsen's spiral

The spiral formed by parametric plot of si , ci is known as Nielsen's spiral. It is also referred to as the Euler spiral, the Cornu spiral, a clothoid, or as a linear-curvature polynomial spiral.

The spiral is also closely related to the Fresnel integrals. This spiral has applications in vision processing, road and track construction and other areas.

Expansion

Various expansions can be used for evaluation of trigonometric integrals, depending on the range of the argument.

Asymptotic Series (for Large Argument)

$$\mathrm{Si}(x) = \frac{\pi}{2} - \frac{\cos x}{x}\left(1 - \frac{2!}{x^2} + \frac{4!}{x^4} - \frac{6!}{x^6}\cdots\right) - \frac{\sin x}{x}\left(\frac{1}{x} - \frac{3!}{x^3} + \frac{5!}{x^5} - \frac{7!}{x^7}\cdots\right)$$

$$\mathrm{Ci}(x) = \frac{\sin x}{x}\left(1 - \frac{2!}{x^2} + \frac{4!}{x^4} - \frac{6!}{x^6}\cdots\right) - \frac{\cos x}{x}\left(\frac{1}{x} - \frac{3!}{x^3} + \frac{5!}{x^5} - \frac{7!}{x^7}\cdots\right).$$

These series are asymptotic and divergent, although can be used for estimates and even precise evaluation at $\Re(x) \gg 1$.

Convergent Series

$$\mathrm{Si}(x) = \sum_{n=0}^{\infty} \frac{(-1)^n x^{2n+1}}{(2n+1)(2n+1)!} = x - \frac{x^3}{3!\cdot 3} + \frac{x^5}{5!\cdot 5} - \frac{x^7}{7!\cdot 7} \pm \cdots$$

$$\mathrm{Ci}(x) = \gamma + \ln x + \sum_{n=1}^{\infty} \frac{(-1)^n x^{2n}}{2n(2n)!} = \gamma + \ln x - \frac{x^2}{2!\cdot 2} + \frac{x^4}{4!\cdot 4} \mp \cdots$$

These series are convergent at any complex x, although for $|x| \gg 1$ the series will converge slowly initially, requiring many terms for high precision.

Relation with the Exponential Integral of Imaginary Argument

The function

$$\mathrm{E}_1(z) = \int_1^{\infty} \frac{\exp(-zt)}{t}\,dt \qquad (\Re(z) \geq 0)$$

is called the exponential integral. It is closely related to Si and Ci,

$$E_1(ix) = i\left(-\frac{\pi}{2} + \mathrm{Si}(x)\right) - \mathrm{Ci}(x) = i\,\mathrm{si}(x) - \mathrm{ci}(x) \qquad (x > 0).$$

As each respective function is analytic except for the cut at negative values of the argument, the area of validity of the relation should be extended to (Outside this range, additional terms which are integer factors of π appear in the expression.)

Cases of imaginary argument of the generalized integro-exponential function are

$$\int_1^\infty \cos(ax)\frac{\ln x}{x}dx = -\frac{\pi^2}{24} + \gamma\left(\frac{\gamma}{2} + \ln a\right) + \frac{\ln^2 a}{2} + \sum_{n\geq 1}\frac{(-a^2)^n}{(2n)!(2n)^2},$$

which is the real part of

$$\int_1^\infty e^{iax}\frac{\ln x}{x}dx = -\frac{\pi^2}{24} + \gamma\left(\frac{\gamma}{2} + \ln a\right) + \frac{\ln^2 a}{2} - \frac{\pi}{2}i(\gamma + \ln a) + \sum_{n\geq 1}\frac{(ia)^n}{n!n^2}.$$

Similarly

$$\int_1^\infty e^{iax}\frac{\ln x}{x^2}dx = 1 + ia[-\frac{\pi^2}{24} + \gamma\left(\frac{\gamma}{2} + \ln a - 1\right) + \frac{\ln^2 a}{2} - \ln a + 1 - \frac{i\pi}{2}(\gamma + \ln a - 1)] + \sum_{n\geq 1}\frac{(ia)^{n+1}}{(n+1)!n^2}.$$

Trigonometric Series

A series in cosines and sines of multiple angles, that is, a series of the form

$$\frac{ao}{2} + \sum_{k=1}^{\infty}(a_k \cos kx + b_k \sin kx)$$

or, in complex form,

$$\sum_{k=-\infty}^{\infty} c_k e^{ikx},$$

where a_k, b_k or, respectively, c_k are called the coefficients of the trigonometric series.

Trigonometric series are first encountered in the work of L. Euler. He obtained the expansions

$$\frac{\pi - x}{2} = \sin x + \frac{\sin 2x}{2} + \frac{\sin 3x}{3} + \ldots \quad (0 < x < 2\pi)$$

$$\frac{1 - r\cos x}{1 - 2r\cos x + r^2} = 1 + r\cos x + r^2\cos 2x + \ldots,$$

$$\frac{r\sin x}{1 - 2r\cos x + r^2} = r\sin x + r^2 \sin 2x + \ldots$$

In the middle of the 18th century, in connection with the study of problems on the free oscillations of strings, there arose the question of the possibility of "representing" functions characterizing the initial position of a string in the form of a sum of a trigonometric series. Functions were usually related to their analytic specifications, which led to the consideration of analytic or piecewise-analytic functions only. But here the need arose to construct a trigonometric series "representing" a function whose graph could be a rather arbitrary curve.

However, the significance of the arguments was even greater. In fact, out of these discussions various questions arose connected with many fundamentally important concepts and ideas of mathematical analysis in general, such as the "representation" of functions by Taylor series and the analytic continuation of functions, the use of divergent series, interchange of limits, infinite systems of equations, interpolation of functions by polynomials, etc.

Subsequently, as well as in this initial period, the theory of trigonometric series served as a source of new ideas of mathematical analysis and influenced the development of other branches of it. Investigations in trigonometric series played an important role in the construction of the Riemann and Lebesgue integrals. The theory of functions of a real variable originated and was then developed in close connection with the theory of trigonometric series. As generalizations of the theory of trigonometric series there emerged the Fourier integral, almost-periodic functions, general orthogonal series, and abstract harmonic analysis. Research into trigonometric series served as a starting point in the creation of set theory. Trigonometric series are a powerful means for representing and studying functions.

The question introduced into the debates of the mathematicians in the 18th century was solved in 1807 by J. Fourier, who gave formulas for calculating the coefficients of the trigonometric series (1) that had to "represent" a function f on $(0,2\pi)$:

$$\left.\begin{aligned} a_k &= \frac{1}{\pi}\int_0^{2\pi} f(x)\cos kx\,dx, \\ b_k &= \frac{1}{\pi}\int_0^{2\pi} f(x)\sin kx\,dx, \end{aligned}\right\}$$

and who applied them in the solution of problems of heat conduction. In above Formulas have acquired the name Fourier formulas, although they were encountered earlier by A. Clairaut and Euler via term-by-term integration. The trigonometric series whose coefficients are defined by above formulas is called the Fourier series of f, and the numbers a_k, b_k the Fourier coefficients of f.

The character of the results obtained depends on how the "representation" of the function by a series is to be understood, and how the integral in is to be understood. The modern form of the theory of trigonometric series was obtained after the appearance of the Lebesgue integral.

The theory of trigonometric series can conditionally be divided into two main branches: the theory of Fourier series, in which it is supposed that the series is the Fourier series of some function, and the theory of general trigonometric series, where this hypothesis is not made. The main results in the theory of general trigonometric series are given below (here the measure of sets and measurability of functions are to be understood in the sense of Lebesgue).

The first systematic study of trigonometric series in which it was not supposed that these series are Fourier series. For this reason, the theory of general trigonometric series is sometimes called the Riemann theory of trigonometric series.

For the study of the properties of an arbitrary series with coefficients converging to zero, Riemann considered the continuous function F that is the sum of the uniformly-convergent series

$$\frac{ao}{4}x^2 - \sum_{k=1}^{\infty} \frac{a_k \cos kx + b_k \sin kx}{k^2},$$

Obtained after twice term-by-term integration of the series. If the series converges at some point x to the numbers, then the second symmetric derivative of F exists at this point and is equal to s:

$$\lim_{h \to 0} \frac{F(x+2h) - 2F(x) + F(x-2h)}{4h^2} = s.$$

Since

$$\frac{F(x+2h) - 2F(x) + F(x-2h)}{4h^2} =$$

$$\frac{ao}{2} + \sum_{k=1}^{\infty} \left(\frac{\sin kh}{kh} \right)^2 (a_k \cos kx + b_k \sin kx),$$

This leads to the summation of produced by the factors $(\sin kh / kh)^2$, called the Riemann summation method. By means of the function F, Riemann formulated the localization principle, according to which the behavior of the series at a point x depends only on the behavior of F in an arbitrarily small neighborhood of this point.

If a trigonometric series converges on a set of positive measure, then its coefficients converge to zero (the Cantor–Lebesgue theorem). Convergence to zero of the coefficients of a trigonometric series also follows from convergence of the series on a set of the second category.

One of the central problems in the theory of general trigonometric series is that of "representing" an arbitrary function by a trigonometric series. By strengthening results of N.N. Luzin on the representation of functions by trigonometric series that are summable almost-everywhere by the methods of Abel–Poisson and Riemann, D.E. Men'shov proved the following theorem, relating to the most important case when the representation of f is understood as convergence of the trigonometric series to f almost everywhere. There exists for any measurable, almost-everywhere finite 2π -periodic function f a trigonometric series converging to it almost-everywhere (Men'shov's theorem). It should be noted that even if the function f is integrable, one cannot in general choose the Fourier series of f for this series, since there exist Fourier series that are everywhere divergent.

Men'shov's theorem can be strengthened as follows: If a 2π -periodic function f is measurable and finite almost everywhere, then there exists a continuous function ϕ such that $\phi' = f$ almost everywhere and the differentiated Fourier series of ϕ converges to f almost everywhere.

It is not known whether the condition that f be finite almost-everywhere can be dropped in

Men'shov's theorem. In particular, it is not known whether there is a trigonometric series converging almost-everywhere to $+\infty$.

Therefore the problem on the representation of functions that can take infinite values on a set of positive measure has been considered for the case when convergence almost-everywhere is replaced by a weaker condition, namely, convergence in measure. Convergence in measure to functions that can take infinite values is defined as follows: The sequence s_n of partial sums of a trigonometric series converges in measure to a function f if

$$s_n(x) = f_n(x) + \alpha_n(x),$$

Where converges to f almost everywhere, while the sequence α_n converges in measure to zero. In this formulation, the question of the representation of functions has been completely solved: there exists for any measurable 2π-periodic function a trigonometric series converging to it in measure.

Much research has been devoted to the problem of uniqueness of trigonometric series: Can two distinct trigonometric series converge to the same function? In another formulation: If a trigonometric series converges to zero, then does it follow that all coefficients of the series are zero? Here one may have in mind convergence at all points or at all points outside some set. The answers to these questions depend essentially on the properties of the set outside which convergence is presupposed.

The following terminology has been established. A set $E \subset [0,2\pi]$ is called a set of uniqueness, or a U-set, if the convergence everywhere on $[0,2\pi]$ of a trigonometric series to zero except, perhaps, at points of E, implies that all coefficients of the series are zero. Otherwise E is called a set of multiplicity, or an M-set. As G. Cantor showed, the empty, as well as any finite, set is a U-set. Any countable set is also a U-set. On the other hand, every set of positive measure is an M-set.

The existence of M-sets of measure zero was established by Men'shov, who constructed the first example of a perfect set possessing these properties. This result is of prime importance in the uniqueness problem. It follows from the existence of M-sets of measure zero that in the representation of functions by trigonometric series converging almost everywhere, these series are automatically non-uniquely determined.

Perfect sets can also be U-sets. In the uniqueness problem, an important role is played by very precise characteristics of sets of measure zero. The general question on the classification of sets of measure zero into M- and U-sets remains open. It has not even been solved for perfect sets.

The following problem touches on the uniqueness problem. If a trigonometric series converges to a function $f \in L(0,2\pi)$, then does this series have to be the Fourier series of f? P. du Bois-Reymond gave an affirmative answer to this question when f is Riemann integrable and the series converges to f at all points. Results of Ch.J. de la Vallée-Poussin imply that the answer is also affirmative in case the series is convergent to a finite sum everywhere except on a countable set of points.

If a trigonometric series converges absolutely at some point , then the points of convergence of the series, as well as the points of its absolute convergence, are distributed symmetrically about x_0.

According to the Denjoy–Luzin theorem, absolute convergence on a set of positive measure of a trigonometric series implies that the series $\sum_k (|a_k| + |b_k|)$ is convergent, and hence that is absolutely convergent for all x. This property is also possessed by sets of the second category and certain sets of measure zero.

This survey only covers one-dimensional trigonometric series. There are separate results relating to general trigonometric series of several variables. Here in many cases it is also necessary to find natural formulations of the problems.

References

- A sentence more appropriate for high schools is "'Some Old Horse Came A''Hopping Through Our Alley". Foster, Jonathan K. (2008). Memory: A Very Short Introduction. Oxford. p. 128. ISBN 0-19-280675-0Trigonometry: byjus.com, Retrieved 16 May 2018
- Rowe, B.; et al. (2015). "GALSIM: The modular galaxy image simulation toolkit". Astronomy and Computing. 10: 121. arXiv:1407.7676 . Bibcode:2015A&C....10..121R. doi:10.1016/j.ascom.2015.02.002
- What-is-Trigonometry: owlcation.com, Retrieved 10 July 2018
- Trigonometric-series: encyclopediaofmath.org, Retrieved 21 April 2018
- Press, WH; Teukolsky, SA; Vetterling, WT; Flannery, BP (2007), "Section 6.8.2. Cosine and Sine Integrals", Numerical Recipes: The Art of Scientific Computing (3rd ed.), New York: Cambridge University Press, ISBN 978-0-521-88068-8
- Trigonometry-and-right-triangles, boundless-algebra: lumenlearning.com, Retrieved 30 March 2018

2

Trigonometric Functions

Trigonometric functions are the functions of an angle which relates the angles of a triangle with the lengths of its sides. This chapter discusses in detail the different trigonometric functions namely the sine function, cosine function, tangent function, cosine function, secant function and cotangent function.

In Mathematics, Trigonometric function is one of six functions (sine, cosine, tangent, cotangent, secant, and cosecant) that represent ratios of sides of right triangles. They are also known as the circular functions, since their values can be defined as ratios of the x and y coordinates of points on a circle of radius 1 that correspond to angles in standard positions. Such values have been tabulated and programmed into scientific calculators and computers. This allows trigonometry to be easily applied to surveying, engineering, and navigation problems in which one of a right triangle's acute angles and the length of a side are known and the lengths of the other sides are to be found. The fundamental trigonometric identity is $\sin^2\theta + \cos^2\theta = 1$, in which θ is an angle. Certain intrinsic qualities of the trigonometric functions make them useful in mathematical analysis. In particular, their derivatives form patterns useful for solving differential equations.

The trigonometric functions relate the angles in a right triangle to the ratios of the sides. Given the following triangle:

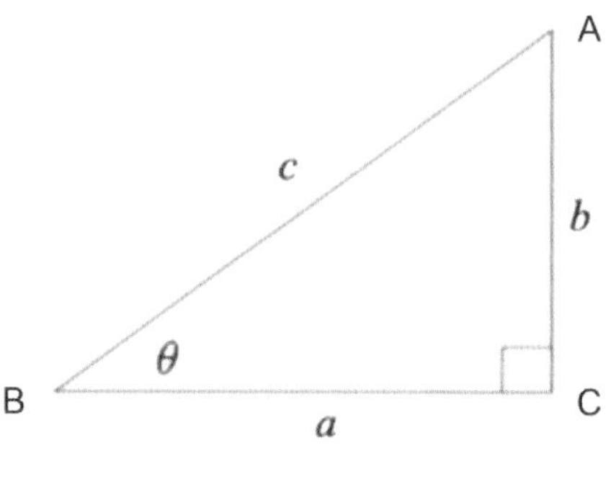

The basic trigonometric functions are defined for $0 < \theta < \frac{\pi}{2}$ as

$$\sin\theta = \frac{b}{c}, \quad \cos\theta = \frac{a}{c}, \quad \tan\theta = \frac{b}{a}.$$

If we consider the angle θ and label the sides with respect to θ, then a is the length of the "adjacent" side, b is the length of the "opposite" side, c and is the length of the hypotenuse. Then the basic trigonometric functions can be expressed as follows:

$$\sin\theta = \frac{\text{opposite}}{\text{hypotenuse}}, \quad \cos\theta = \frac{\text{adjacent}}{\text{hypotenuse}}, \quad \tan\theta = \frac{\text{opposite}}{\text{adjacent}}.$$

However, a more useful definition comes from the unit circle. If we consider a circle with a radius of 1 unit, centered at the origin, then the angle θ inside the circle describes a right triangle when we drop a perpendicular to the x-axis from the point of intersection with the circle.

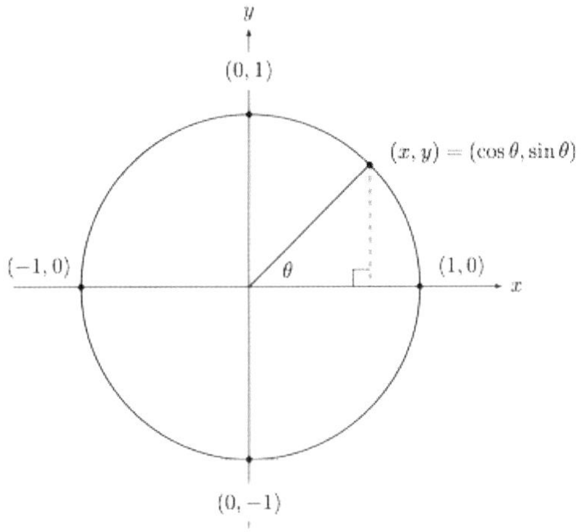

Unit Circle

Notice that the right triangle so described has a hypotenuse equal to the radius of the circle, an adjacent side equal to the x-coordinate of the point (x, y) and an opposite side equal to the y-coordinate. This gives rise naturally to the following refined definitions:

$$\sin\theta = \frac{y}{\text{radius}}, \qquad \cos\theta = \frac{x}{\text{radius}}, \qquad \tan\theta = \frac{y}{x}.$$

As shown in the above diagram, since the radius is 1 in the unit circle, this simplifies to $x = \cos\theta$ and $y = \sin\theta$.

These definitions have the advantage of being compatible with the triangle definition above, as well as allowing the evaluation of angles corresponding to any real number.

There are certain values of these functions which are useful to remember. They are:

θ	0°	$\frac{\pi}{6} = 30^{\circ}$	$\frac{\pi}{4} = 45^{\circ}$	$\frac{\pi}{3} = 60^{\circ}$	$\frac{\pi}{2} = 90^{\circ}$
$\sin\theta$	$\frac{\sqrt{0}}{2}$	$\frac{\sqrt{1}}{2}$	$\frac{\sqrt{2}}{2}$	$\frac{\sqrt{3}}{2}$	$\frac{\sqrt{4}}{2}$
$\cos\theta$	$\frac{\sqrt{4}}{2}$	$\frac{\sqrt{3}}{2}$	$\frac{\sqrt{2}}{2}$	$\frac{\sqrt{1}}{2}$	$\frac{\sqrt{0}}{2}$
$\tan\theta$	0	$\frac{1}{\sqrt{3}}$	1	$\sqrt{3}$	∞

The reason for writing them in this way is to aid remembering these terms. For example, the numerator for $\sin\theta$ is simply the square root of 0,1,2,3, or 4.

Example

What values of θ satisfy

$$0 < \theta < 2\pi \quad \text{and} \quad \cos\theta = 0?$$

A good approach for a problem like this is to imagine the unit circle diagram shown above. In this unit circle diagram, $\cos\theta$ is the x coordinate. Thus, the problem is asking for angle θ whose x -coordinate is equal to y ,which occurs at the two points where the unit circle intersects the y -axis: $\theta = \frac{\pi}{2}$ and $\theta = \frac{3\pi}{2}$. Since these angles satisfy the given conditions, the two possible angles are $\theta = \frac{\pi}{2}$ and $\theta = \frac{3\pi}{2}$.

Example

If θ is an angle such that $\cos\theta = 0$, what are the possible values of $\sin\theta$?

Solution 1:

As we saw above, $\cos\theta = 0$, corresponds to points on the unit circle whose x-coordinate is 0. Since these points occur at the points of intersection with the y-axis, the possible values of $\sin\theta$ are the possible y-coordinate which are 1 and -1.

Solution 2:

From the first example, if $\cos\theta = 0$, and $0 \le \theta < 2\pi$, then $\theta = \frac{\pi}{2}$ or $\theta = \frac{3\pi}{2}$. Note that for all other values of θ outside of this range, we have $\sin\theta = \sin(\theta \pm 2\pi)$, so we can add or subtract multiples of 2π until θ lies in this range. For $\theta = \frac{\pi}{2}$, we have $\sin\theta = 1$ and for $\theta = \frac{3\pi}{2}$. we have $\sin\theta = -1$. Therefore, the possible values of $\sin\theta$ are 1 and -1.

Example

If θ is an angle in a right triangle such that $\sin\theta = \frac{1}{2}$, what is the value of $\cos\theta$?

Since θ is an angle in a right triangle, it must be the case that $0 < \theta < \frac{\pi}{2}$.

From the table above, we see that the value θ such that $\sin(\theta) = \frac{1}{2}$ is $\theta = \frac{\pi}{6} = 30^\circ$. Also, since the y -coordinate of the points on the unit circle are increasing as θ goes from 0 to $\frac{\pi}{2}$, the value $\theta = \frac{\pi}{6}$ is the unique value such that $0 < \theta < \frac{\pi}{2}$ and $\sin\theta = \frac{1}{2}$ Then, we have $\cos\theta = \cos\frac{\pi}{2} = \frac{\sqrt{3}}{2}$.

Example

In the table of specific values for trigonometric functions above, why is there no value for $\tan\frac{\pi}{2}$? What happens to $\tan\theta$ as θ gets closer and closer to $\frac{\pi}{2}$?

From the above definition of $\tan\theta$ we have $\tan\theta = \frac{\sin\theta}{\cos\theta} = \frac{y}{x}$, where (x,y) are the x- and y-coordinates of the point of angle θ on the unit circle. As θ moves towards $\frac{\pi}{2}$, $\cos\theta$ (the x-coordinate) becomes smaller and smaller, whereas $\sin\theta$ (the y-coordinate) becomes closer and closer to 1.Therefore, the numerator of $\tan\theta = \frac{y}{x}$ approaches 1 and the denominator approaches 0, implying $\tan\theta$ approaches infinity.

Specific Values - Basic

There are certain values of the basic trigonometric functions which are useful to remember. They are:

θ	0°	$\frac{\pi}{6} = 30^{\circ}$	$\frac{\pi}{4} = 45^{\circ}$	$\frac{\pi}{3} = 60^{\circ}$	$\frac{\pi}{2} = 90^{\circ}$
$\sin\theta$	$\frac{\sqrt{0}}{2}$	$\frac{\sqrt{1}}{2}$	$\frac{\sqrt{2}}{2}$	$\frac{\sqrt{3}}{2}$	$\frac{\sqrt{4}}{2}$
$\cos\theta$	$\frac{\sqrt{4}}{2}$	$\frac{\sqrt{3}}{2}$	$\frac{\sqrt{2}}{2}$	$\frac{\sqrt{1}}{2}$	$\frac{\sqrt{0}}{2}$
$\tan\theta$	0	$\frac{1}{\sqrt{3}}$	1	$\sqrt{3}$	$\pm\infty$

The reason for writing them in this way is to aid remembering these terms. For example, the numerator for $\sin\theta$ is simply the square root of 0, 1, 2, 3, 4.

Visualization in Unit Circle

We can also visualize the cosine and sine values for these values in the unit circle:

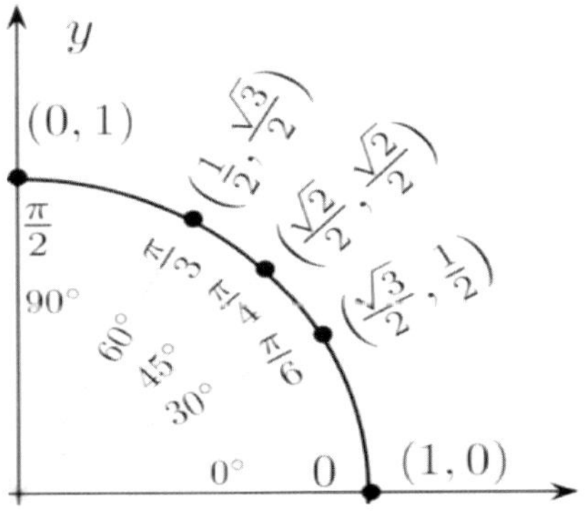

Since the cosine function corresponds to x values, the cosine function will be positive when x values are positive and will be negative when x values are negative. Similarly, since the sine function corresponds to y values, the sine function will be positive when y values are positive and will be negative when y values are negative. This gives us the following behavior in the four quadrants of the plane:

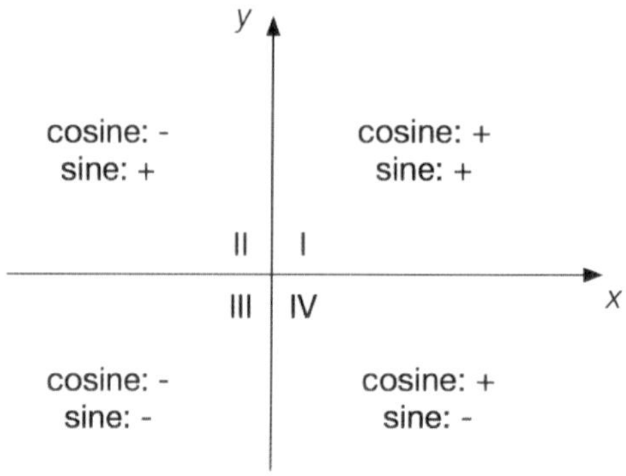

Then by using the few specific values from the first quadrant, we can figure out the specific values of the cosine and sine functions in all quadrants. Here is the visualization for all quadrants:

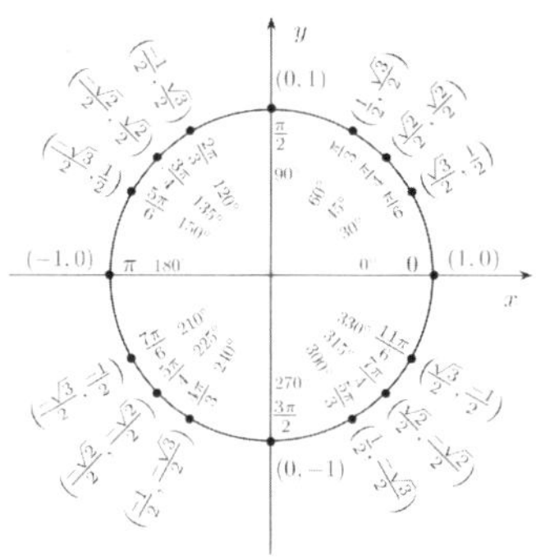

Example

What are the values of θ in the range $0 \le \theta < 2\pi$ such that $\sin\theta - \cos\theta$?

From the unit circle visualization above, we see that $\theta = \frac{\pi}{4}$ and $\theta = \frac{5\pi}{4}$ satisfy

$$\sin\left(\frac{\pi}{4}\right) = \cos\left(\frac{\pi}{4}\right) = \frac{\sqrt{2}}{2}$$

$$\sin\left(\frac{5\pi}{4}\right) = \cos\left(\frac{5\pi}{4}\right) = -\frac{\sqrt{2}}{2}.$$

We also observe that the line $y = x$ intersects the unit circles for only these two values of θ, so the values of θ satisfying the required conditions are $\theta = \frac{\pi}{4}$ and $\theta = \frac{5\pi}{4}$.

Example

What values of θ in the range $0 \le \theta < 2\pi$ satisfy $\sin\theta \ge \frac{\sqrt{2}}{2}$?

By drawing the line $y = \frac{\sqrt{2}}{2}$, we would like to find the values of θ such that the the y-value of the angle θ on the unit circle lies above this line (since $\sin\theta$ corresponds to the y-coordinate of the unit circle). This holds for $\theta \in \left[\frac{\pi}{4}, \frac{3\pi}{4}\right]$, so these are the values of θ satisfying $\sin\theta \geq \frac{\sqrt{2}}{2}$.

Example

What are the values of θ in the range $0 \leq \theta < 2\pi$ such that $\sin\theta = -\cos\theta$?

From the unit circle visualization above, we see that $\theta = \frac{3\pi}{4}$ and $\theta = \frac{7\pi}{4}$ satisfy

$$\sin\left(\frac{3\pi}{4}\right) = -\cos\left(\frac{3\pi}{4}\right) = \frac{\sqrt{2}}{2}$$

$$\sin\left(\frac{7\pi}{4}\right) = -\cos\left(\frac{7\pi}{4}\right) = -\frac{\sqrt{2}}{2}.$$

We also observe that the line $y = -x$ intersects the unit circle for only these two values of θ, so the values of θ satisfying the required conditions are $\theta = \frac{3\pi}{4}$ and $\theta = \frac{7\pi}{4}$.

Algebraic Values

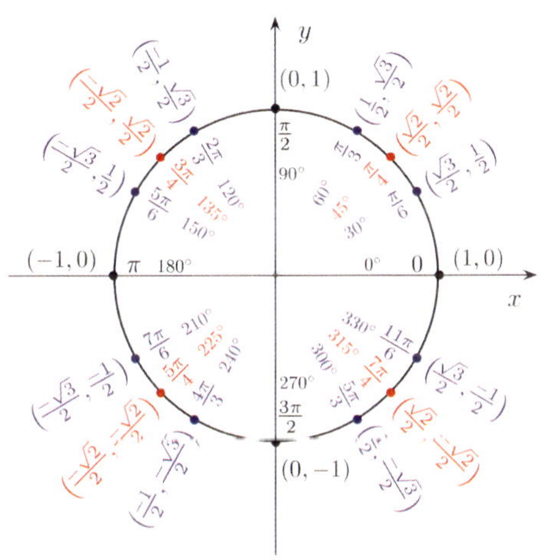

The unit circle, with some points labeled with their cosine and sine (in this order), and the corresponding angles in radians and degrees.

The algebraic expressions for sin(0°), sin(30°), sin(45°), sin(60°) and sin(90°) are

$$0, \quad \frac{1}{2}, \quad \frac{\sqrt{2}}{2}, \quad \frac{\sqrt{3}}{2}, \quad 1,$$

respectively. Writing the numerators as square roots of consecutive natural numbers $\left(\frac{\sqrt{0}}{2}, \frac{\sqrt{1}}{2}, \frac{\sqrt{2}}{2}, \frac{\sqrt{3}}{2}, \frac{\sqrt{4}}{2}\right)$ provides an easy way to remember the values. Such simple expressions generally do not exist for other angles which are rational multiples of a straight angle.

For an angle which, measured in degrees, is a multiple of three, the sine and the cosine may be expressed in terms of square roots, as shown below. These values of the sine and the cosine may thus be constructed by ruler and compass.

For an angle of an integer number of degrees, the sine and the cosine may be expressed in terms of square roots and the cube root of a non-real complex number. Galois theory allows proof that, if the angle is not a multiple of 3°, non-real cube roots are unavoidable.

For an angle which, measured in degrees, is a rational number, the sine and the cosine are algebraic numbers, which may be expressed in terms of *n*th roots. This results from the fact that the Galois groups of the cyclotomic polynomials are cyclic.

For an angle which, measured in degrees, is not a rational number, then either the angle or both the sine and the cosine are transcendental numbers. This is a corollary of Baker's theorem, proved in 1966.

Explicit Values

Algebraic expressions for 15°, 18°, 36°, 54°, 72° and 75° are as follows:

$$\sin 15^\circ = \cos 75^\circ = \frac{\sqrt{6}-\sqrt{2}}{4}$$

$$\sin 18^\circ = \cos 72^\circ = \frac{\sqrt{5}-1}{4}$$

$$\sin 36^\circ = \cos 54^\circ = \frac{\sqrt{10-2\sqrt{5}}}{4}$$

$$\sin 54^\circ = \cos 36^\circ = \frac{\sqrt{5}+1}{4}$$

$$\sin 72^\circ = \cos 18^\circ = \frac{\sqrt{10+2\sqrt{5}}}{4}$$

$$\sin 75^\circ = \cos 15^\circ = \frac{\sqrt{6}+\sqrt{2}}{4}.$$

From these, the algebraic expressions for all multiples of 3° can be computed. For example:

$$\sin 3^\circ = \cos 87^\circ = \frac{2\left(1-\sqrt{3}\right)\sqrt{5+\sqrt{5}}+\left(1+\sqrt{3}\right)\left(\sqrt{10}-\sqrt{2}\right)}{16}$$

$$\sin 6^\circ = \cos 84^\circ = \frac{\sqrt{30-6\sqrt{5}}-\sqrt{5}-1}{8}$$

$$\sin 9^\circ = \cos 81^\circ = \frac{\sqrt{10}+\sqrt{2}-2\sqrt{5-\sqrt{5}}}{8}$$

$$\sin 84^\circ = \cos 6^\circ = \frac{\sqrt{10-2\sqrt{5}}+\sqrt{3}+\sqrt{15}}{8}$$

$$\sin 87^\circ = \cos 3^\circ = \frac{2\left(1+\sqrt{3}\right)\sqrt{5+\sqrt{5}}-\left(1-\sqrt{3}\right)\left(\sqrt{10}-\sqrt{2}\right)}{16}.$$

Algebraic expressions can be deduced for other angles of an integer number of degrees, for example,

$$\sin 1^\circ = \frac{\sqrt[3]{z}-\frac{1}{\sqrt[3]{z}}}{2i},$$

where $z = a + ib$, and a and b are the above algebraic expressions for, respectively, cos 3° and sin 3°, and the principal cube root (that is, the cube root with the largest real part) is to be taken.

Series Definitions

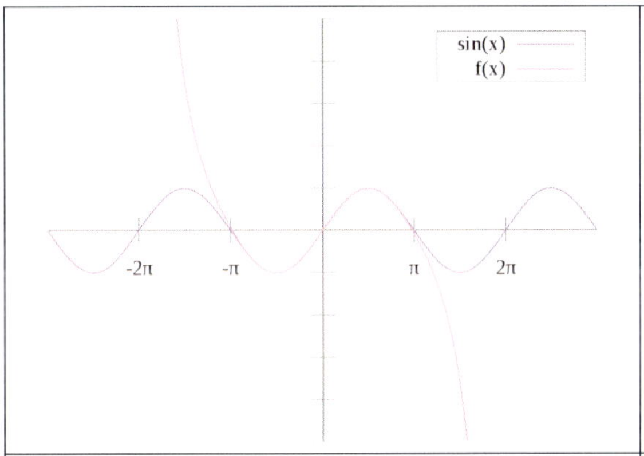	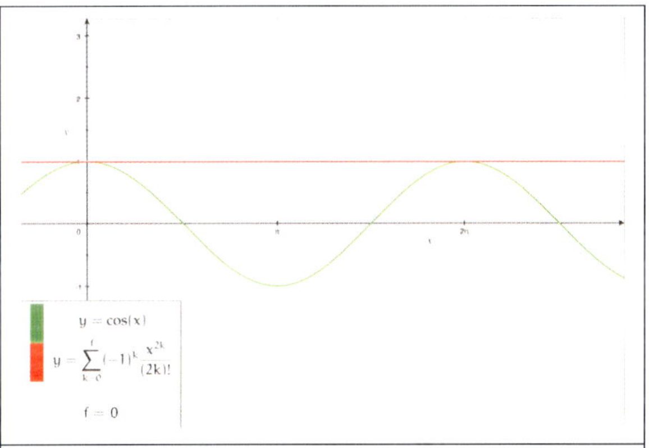
The sine function (blue) is closely approximated by its Taylor polynomial of degree 7 (pink) for a full cycle centered on the origin.	figure for the approximation of cosine via Taylor polynomials.

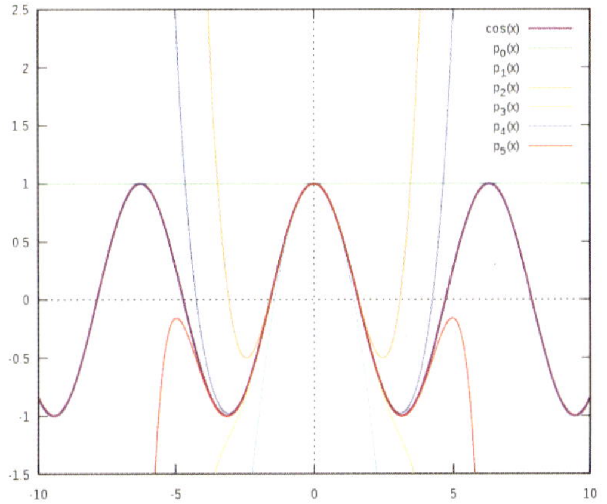

cos(x) together with the first Taylor polynomials $p_n(x) = \sum_{k=0}^{n} (-1)^k \frac{x^{2k}}{(2k)!}$

Trigonometric functions are analytic functions. Using only geometry and properties of limits, it can be shown that the derivative of sine is cosine and the derivative of cosine is the negative of sine. One can then use the theory of Taylor series to show that the following identities hold for all real numbers x. Here, and generally in calculus, all angles are measured in radians.

$$\sin x = x - \frac{x^3}{3!} + \frac{x^5}{5!} - \frac{x^7}{7!} + \cdots$$

$$= \sum_{n=0}^{\infty} \frac{(-1)^n x^{2n+1}}{(2n+1)!}$$

$$\cos x = 1 - \frac{x^2}{2!} + \frac{x^4}{4!} - \frac{x^6}{6!} + \cdots$$

$$= \sum_{n=0}^{\infty} \frac{(-1)^n x^{2n}}{(2n)!}$$

The infinite series appearing in these identities are convergent in the whole complex plane and are often taken as the definitions of the sine and cosine functions of a complex variable. Another standard (and equivalent) definition of the sine and the cosine as functions of a complex variable is through their differential equation, below.

Other series can be found. For the following trigonometric functions:

U_n is the nth up/down number,

B_n is the nth Bernoulli number, and

E_n (below) is the nth Euler number.

$$\tan x = \sum_{n=0}^{\infty} \frac{U_{2n+1} x^{2n+1}}{(2n+1)!}$$

$$= \sum_{n=1}^{\infty} \frac{(-1)^{n-1} 2^{2n} (2^{2n} - 1) B_{2n} x^{2n-1}}{(2n)!}$$

$$= x + \frac{1}{3}x^3 + \frac{2}{15}x^5 + \frac{17}{315}x^7 + \cdots, \qquad \text{for } |x| < \frac{\pi}{2}.$$

$$\csc x = \sum_{n=0}^{\infty} \frac{(-1)^{n+1} 2(2^{2n-1} - 1) B_{2n} x^{2n-1}}{(2n)!}$$

$$= x^{-1} + \frac{1}{6}x + \frac{7}{360}x^3 + \frac{31}{15120}x^5 + \cdots, \qquad \text{for } 0 < |x| < \pi.$$

$$\sec x = \sum_{n=0}^{\infty} \frac{U_{2n} x^{2n}}{(2n)!} = \sum_{n=0}^{\infty} \frac{(-1)^n E_{2n} x^{2n}}{(2n)!}$$

$$= 1 + \frac{1}{2}x^2 + \frac{5}{24}x^4 + \frac{61}{720}x^6 + \cdots, \qquad \text{for } |x| < \frac{\pi}{2}.$$

$$\cot x = \sum_{n=0}^{\infty} \frac{(-1)^n 2^{2n} B_{2n} x^{2n-1}}{(2n)!}$$
$$= x^{-1} - \frac{1}{3}x - \frac{1}{45}x^3 - \frac{2}{945}x^5 - \cdots, \qquad \text{for } 0 < |x| < \pi.$$

When the series for the tangent and secant functions are expressed in a form in which the denominators are the corresponding factorials, the numerators, called the "tangent numbers" and "secant numbers" respectively, have a combinatorial interpretation: they enumerate alternating permutations of finite sets, of odd cardinality for the tangent series and even cardinality for the secant series. The series itself can be found by a power series solution of the aforementioned differential equation.

From a theorem in complex analysis, there is a unique analytic continuation of this real function to the domain of complex numbers. They have the same Taylor series, and so the trigonometric functions are defined on the complex numbers using the Taylor series above.

There is a series representation as partial fraction expansion where just translated reciprocal functions are summed up, such that the poles of the cotangent function and the reciprocal functions match:

$$\pi \cdot \cot(\pi x) = \lim_{N \to \infty} \sum_{n=-N}^{N} \frac{1}{x+n}.$$

This identity can be proven with the Herglotz trick. Combining the $(-n)$th with the nth term lead to absolutely convergent series:

$$\pi \cdot \cot(\pi x) = \frac{1}{x} + \sum_{n=1}^{\infty} \frac{2x}{x^2 - n^2}, \qquad \frac{\pi}{\sin(\pi x)} = \frac{1}{x} + \sum_{n=1}^{\infty} \frac{(-1)^n 2x}{x^2 - n^2}.$$

Relationship to Exponential Function and Complex Numbers

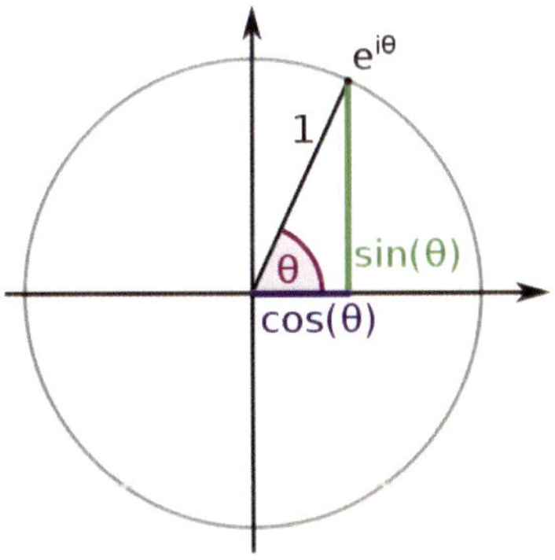

$\cos(\theta)$ and $\sin(\theta)$ are the real and imaginary part of $e^{i\theta}$ respectively.

It can be shown from the series definitions that the sine and cosine functions are respectively the imaginary and real parts of the exponential function of a purely imaginary argument. That is, if x is real, we have

$$\cos x = \operatorname{Re}(e^{ix}), \qquad \sin x = \operatorname{Im}(e^{ix}),$$

and

$$e^{ix} = \cos x + i \sin x.$$

The latter identity, although primarily established for real x, remains valid for every complex x, and is called Euler's formula.

Euler's formula can be used to derive most trigonometric identities from the properties of the exponential function, by writing sine and cosine as:

$$\sin x = \frac{e^{ix} - e^{-ix}}{2i}, \qquad \cos x = \frac{e^{ix} + e^{-ix}}{2}.$$

It is also sometimes useful to express the complex sine and cosine functions in terms of the real and imaginary parts of their arguments.

$$\sin(x + iy) = \sin x \cosh y + i \cos x \sinh y,$$
$$\cos(x + iy) = \cos x \cosh y - i \sin x \sinh y.$$

This exhibits a deep relationship between the complex sine and cosine functions and their real (sin, cos) and hyperbolic real (sinh, cosh) counterparts.

Complex Graphs

In the following graphs the domain is the complex plane pictured with domain coloring, and the range values are indicated at each point by color. Brightness indicates the size (absolute value) of the range value, with black being zero. Hue varies with argument, or angle, measured from the positive real axis.

Trigonometric functions in the complex plane					
		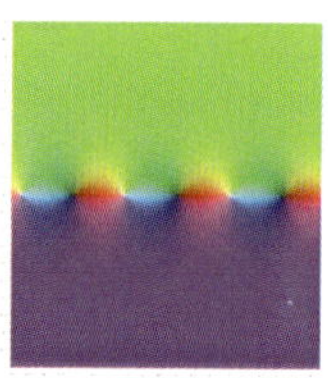	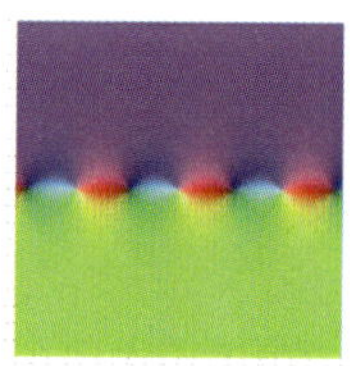	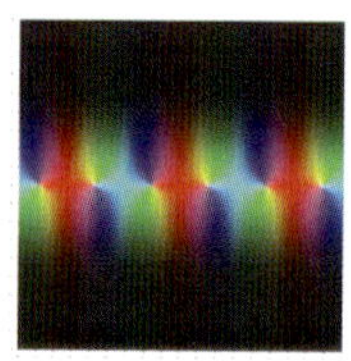	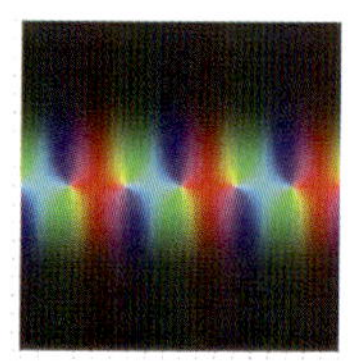
$\sin z$	$\cos z$	$\tan z$	$\cot z$	$\sec z$	$\csc z$

Definitions via Differential Equations

Both the sine and cosine functions satisfy the linear differential equation:

$$y'' = -y$$

That is to say, each is the additive inverse of its own second derivative. Within the 2-dimensional function space V consisting of all solutions of this equation,

- The sine function is the unique solution satisfying the initial condition $(y'(0), y(0)) = (1, 0)$.
- The cosine function is the unique solution satisfying the initial condition $(y'(0), y(0)) = (0, 1)$.

Since the sine and cosine functions are linearly independent, together they form a basis of V. This method of defining the sine and cosine functions is essentially equivalent to using Euler's formula. It turns out that this differential equation can be used not only to define the sine and cosine functions but also to prove the trigonometric identities for the sine and cosine functions.

Further, the observation that sine and cosine satisfies $y'' = -y$ means that they are eigenfunctions of the second-derivative operator.

The tangent function is the unique solution of the nonlinear differential equation

$$y' = 1 + y^2$$

Satisfying the initial condition $y(0) = 0$. There is a very interesting visual proof that the tangent function satisfies this differential equation.

The Significance of Radians

Radians specify an angle by measuring the length around the path of the unit circle and constitute a special argument to the sine and cosine functions. In particular, only sines and cosines that map radians to ratios satisfy the differential equations that classically describe them. If an argument to sine or cosine in radians is scaled by frequency,

$$f(x) = \sin(kx),$$

then the derivatives will scale by amplitude.

$$f'(x) = k\cos(kx).$$

Here, k is a constant that represents a mapping between units. If x is in degrees, then

$$k = \frac{\pi}{180^\circ}.$$

This means that the second derivative of a sine in degrees does not satisfy the differential equation

$$y'' = -y$$

but rather

$$y'' = -k^2 y.$$

The cosine's second derivative behaves similarly.

This means that these sines and cosines are different functions, and that the fourth derivative of sine will be sine again only if the argument is in radians.

Inverse Functions

The trigonometric functions are periodic, and hence not injective, so strictly they do not have an inverse function. Therefore, to define an inverse function we must restrict their domains so that the trigonometric function is bijective. In the following, the functions on the left are *defined* by the equation on the right; these are not proved identities. The principal inverses are usually defined as:

Function	Definition	Value Field
$\arcsin x = y$	$\sin y = x$	$-\frac{\pi}{2} \le y \le \frac{\pi}{2}$
$\arccos x = y$	$\cos y = x$	$0 \le y \le \pi$
$\arctan x = y$	$\tan y = x$	$-\frac{\pi}{2} < y < \frac{\pi}{2}$
$\operatorname{arccot} x = y$	$\cot y = x$	$0 < y < \pi$
$\operatorname{arcsec} x = y$	$\sec y = x$	$0 \le y \le \pi, y \ne \frac{\pi}{2}$
$\operatorname{arccsc} x = y$	$\csc y = x$	$-\frac{\pi}{2} \le y \le \frac{\pi}{2}, y \ne 0$

The notations $\sin^{-1}$ and $\cos^{-1}$ are often used for arcsin and arccos, etc. When this notation is used, the inverse functions could be confused with the multiplicative inverses of the functions. The notation using the "arc-" prefix avoids such confusion, though "arcsec" for arcsecant can be confused with "arcsecond".

Just like the sine and cosine, the inverse trigonometric functions can also be defined in terms of infinite series. For example,

$$\arcsin z = z + \left(\frac{1}{2}\right)\frac{z^3}{3} + \left(\frac{1 \cdot 3}{2 \cdot 4}\right)\frac{z^5}{5} + \left(\frac{1 \cdot 3 \cdot 5}{2 \cdot 4 \cdot 6}\right)\frac{z^7}{7} + \cdots.$$

These functions may also be defined by proving that they are antiderivatives of other functions. The arcsine, for example, can be written as the following integral:

$$\arcsin z = \int_0^z \frac{1}{\sqrt{1-x^2}}\,dx, \quad |z| < 1.$$

Analogous formulas for the other functions can be found at inverse trigonometric functions. Using the complex logarithm, one can generalize all these functions to complex arguments:

$$\arcsin z = -i \log\left(iz + \sqrt{1-z^2}\right),$$

$$\arccos z = -i \log\left(z + \sqrt{z^2 - 1}\right),$$

$$\arctan z = \frac{1}{2} i \log\left(\frac{1-iz}{1+iz}\right).$$

Connection to the Inner Product

In an inner product space, the angle between two non-zero vectors is defined to be

$$\operatorname{angle}(x, y) = \arccos \frac{\langle x, y \rangle}{\|x\| \cdot \|y\|}.$$

Sine Function

The trigonometric *sine function* is closely related to the Greek *chord function*. In fact the sine function, sometimes known as the *half-chord function*, has the same relationship with the *radius* of a circle as the chord function has with the *diameter* of a circle.

A definition of the sine as the relationship between one half of a chord and one half of the central angle subtended by the chord. It also included tables of sine values from *zero degrees* (0°) up to *ninety degrees* (90°) in twenty-four increments of *three and a quarter degrees* (3.75°). They are believed to be the oldest surviving examples of sine tables, and are accurate to four decimal places. The relationship between the half-chord and the central angle θ that it subtends is illustrated below.

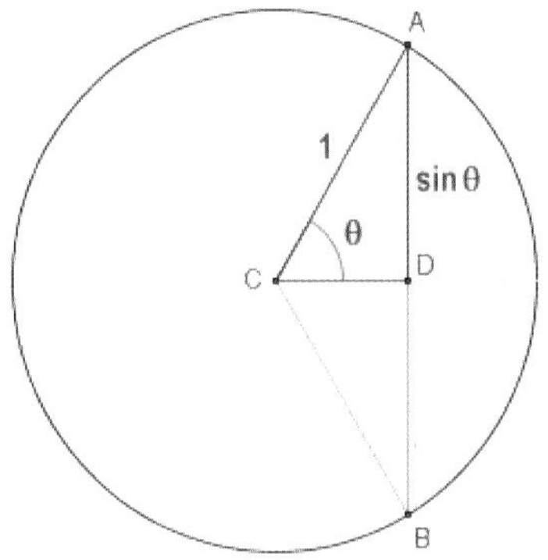

Angle θ is the central angle subtended by half-chord *AD*

As you can hopefully see from the diagram, the half-chord *AD* is one leg of right-angled triangle ADC. The other leg is line segment *CD*, which is the *perpendicular bisector* of chord *AB*. The hypotenuse *AC* of the right-angled triangle is a radius of the circle. We have used a *unit circle* here (i.e. a circle with a radius of *one unit*), which means that by definition, half-chord *AD* is equal to the sine of angle θ. For a circle of arbitrary radius r, the length of half-chord *AD* would be equal to $sin(\theta)r$. The relationship between the chord function and the sine function is given by the equation:

$$crd(\alpha) = 2sin\left(\frac{\alpha}{2}\right) = 2sin(\theta)$$

where α is the central angle subtended by the chord. Just as the chord function used by Greek astronomers and mathematicians is all about the properties of a right-angled triangle inscribed in a *semi-circle*.

The way in which the sine function relates to the right angled triangle is illustrated below. In any right-angled triangle, one of the internal angles will always be *ninety degrees* (90°). Since the internal angles in any triangle must always sum to *one hundred and eighty degrees* (180°), the other two internal angles must both be *acute angles* (i.e. less than ninety degrees). The side opposite the right-angle is always the longest side, and is called the *hypotenuse*. In relation to either one of the acute angles, the side opposite the angle is referred to as the *opposite*. The remaining side (the side common to both the angle in which we are interested and the right angle) is called the *adjacent*.

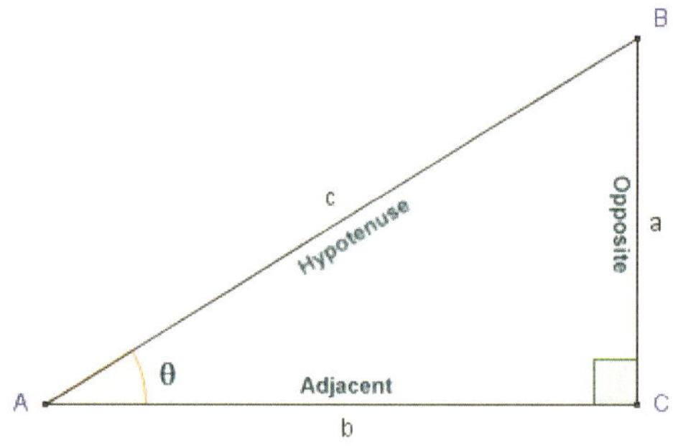

In triangle ABC, $sin(\theta) = a/c$

In triangle ABC, we are interested in angle θ. We have used the convention of labelling each vertex with an upper case character, and the side *opposite* each vertex with the corresponding lower-case character. The *opposite* in relation to angle θ is side a. The *adjacent* is side b, and the *hypotenuse* is side c. The *sine* of an angle in a right-angled triangle is defined as the quotient of the lengths of the opposite and the hypotenuse. Thus, in triangle ABC, the sine of angle θ is given as:

$$sin(\theta) = \frac{a}{c}$$

In the illustration below, and using the same right-angled triangle, we show part of a circle with its center at point A and a radius $r = c$. We have extended line segment BC to a point D on the perimeter of the circle, so that line segment BD becomes a *chord* of the circle. In relation to triangle ABC, side a (the *opposite*) becomes the *half-cord* subtending angle θ. Side b (the *adjacent*) becomes the chord's *perpendicular bisector*, and angle θ is half the central angle subtended by the chord. We include this diagram to emphasize the relationship between right-angled triangles and circles.

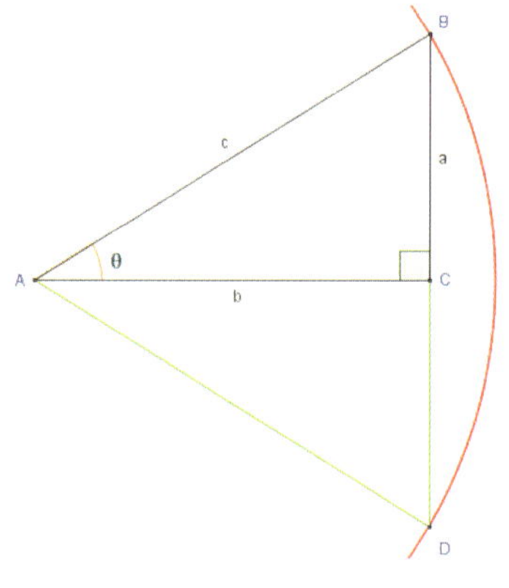

Angle θ is half of the central angle subtended by chord BD

Identities

Exact identities (using radians):

These apply for all values of θ.

$$\sin(\theta) = \cos\left(\frac{\pi}{2} - \theta\right) = \frac{1}{\csc(\theta)}$$

Reciprocal

The reciprocal of sine is cosecant, i.e., the reciprocal of sin(*A*) is csc(*A*), or cosec(*A*). Cosecant gives the ratio of the length of the hypotenuse to the length of the opposite side:

$$\csc(A) = \frac{1}{\sin(A)} = \frac{\text{hypotenuse}}{\text{opposite}} = \frac{h}{a}.$$

Inverse

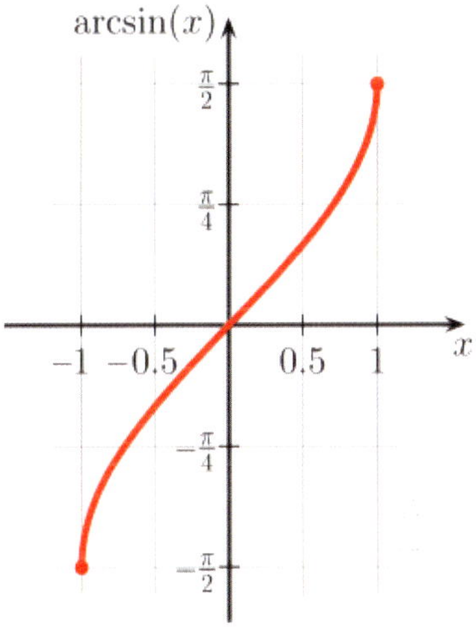

The usual principal values of the arcsin (*x*) function graphed on the cartesian plane. Arcsin is the inverse of sin.

The inverse function of sine is arcsine (arcsin or asin) or inverse sine ($\sin^{-1}$). As sine is non-injective, it is not an exact inverse function but a partial inverse function. For example, sin(0) = 0, but also sin(π) = 0, sin(2π) = 0 etc. It follows that the arcsine function is multivalued: arcsin(0) = 0, but also arcsin(0) = π, arcsin(0) = 2π, etc. When only one value is desired, the function may be restricted to its principal branch. With this restriction, for each *x* in the domain the expression arcsin(*x*) will evaluate only to a single value, called its principal value.

$$\theta = \arcsin\left(\frac{\text{opposite}}{\text{hypotenuse}}\right) = \sin^{-1}\left(\frac{a}{h}\right).$$

k is some integer:

$$\sin(y) = x \Leftrightarrow y = \arcsin(x) + 2\pi k, \text{ or}$$
$$y = \pi - \arcsin(x) + 2\pi k$$

Or in one equation:

$$\sin(y) = x \Leftrightarrow y = (-1)^k \arcsin(x) + \pi k$$

Arcsin satisfies:

$$\sin(\arcsin(x)) = x$$

and

$$\arcsin(\sin(\theta)) = \theta \quad \text{for } -\frac{\pi}{2} \le \theta \le \frac{\pi}{2}.$$

Calculus

For the sine function:

$$f(x) = \sin(x)$$

The derivative is:

$$f'(x) = \cos(x)$$

The antiderivative is:

$$\int f(x)dx = -\cos(x) + C$$

C denotes the constant of integration.

Other Trigonometric Functions

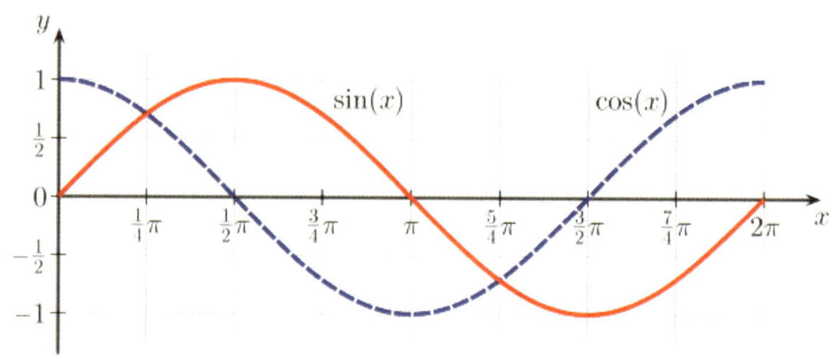

The sine and cosine functions are related in multiple ways. The two functions are out of phase by 90°: $\sin(\pi/2 - x) = \cos(x)$ for all angles x. Also, the derivative of the function $\sin(x)$ is $\cos(x)$.

It is possible to express any trigonometric function in terms of any other (up to a plus or minus sign, or using the sign function).

Sine in terms of the other common trigonometric functions:

	$f\,\theta$	Using plus/minus (±)					Using sign function (sgn)
		$f\,\theta =$	± per Quadrant				$f\,\theta =$
			I	II	III	IV	
cos	$\sin(\theta)$	$= \pm\sqrt{1-\cos^2(\theta)}$	+	+	−	−	$= \operatorname{sgn}\left(\cos\left(\theta - \frac{\pi}{2}\right)\right)\sqrt{1-\cos^2(\theta)}$
	$\cos(\theta)$	$= \pm\sqrt{1-\sin^2(\theta)}$	+	−	−	+	$= \operatorname{sgn}\left(\sin\left(\theta + \frac{\pi}{2}\right)\right)\sqrt{1-\sin^2(\theta)}$

cot	sin(θ)	$=\pm\frac{1}{\sqrt{1+\cot^2(\theta)}}$	+	+	−	−	$=\operatorname{sgn}\left(\cot\left(\frac{\theta}{2}\right)\right)\frac{1}{\sqrt{1+\cot^2(\theta)}}$
	cot(θ)	$=\pm\frac{\sqrt{1-\sin^2(\theta)}}{\sin(\theta)}$	+	−	−	+	$=\operatorname{sgn}\left(\sin\left(\theta+\frac{\pi}{2}\right)\right)\frac{\sqrt{1-\sin^2(\theta)}}{\sin(\theta)}$
tan	sin(θ)	$=\pm\frac{\tan(\theta)}{\sqrt{1+\tan^2(\theta)}}$	+	−	−	+	$=\operatorname{sgn}\left(\tan\left(\frac{2\theta+\pi}{4}\right)\right)\frac{\tan(\theta)}{\sqrt{1+\tan^2(\theta)}}$
	tan(θ)	$=\pm\frac{\sin(\theta)}{\sqrt{1-\sin^2(\theta)}}$	+	−	−	+	$=\operatorname{sgn}\left(\sin\left(\theta+\frac{\pi}{2}\right)\right)\frac{\sin(\theta)}{\sqrt{1-\sin^2(\theta)}}$
sec	sin(θ)	$=\pm\frac{\sqrt{\sec^2(\theta)-1}}{\sec(\theta)}$	+	−	+	−	$=\operatorname{sgn}\left(\sec\left(\frac{4\theta-\pi}{2}\right)\right)\frac{\sqrt{\sec^2(\theta)-1}}{\sec(\theta)}$
	sec(θ)	$=\pm\frac{1}{\sqrt{1-\sin^2(\theta)}}$	+	−	−	+	$=\operatorname{sgn}\left(\sin\left(\theta+\frac{\pi}{2}\right)\right)\frac{1}{\sqrt{1-\sin^2(\theta)}}$

Note that for all equations which use plus/minus (±), the result is positive for angles in the first quadrant.

The basic relationship between the sine and the cosine can also be expressed as the Pythagorean trigonometric identity:

$$\cos^2(\theta)+\sin^2(\theta)=1$$

where $\sin^2(x)$ means $(\sin(x))^2$.

Properties Relating to the Quadrants

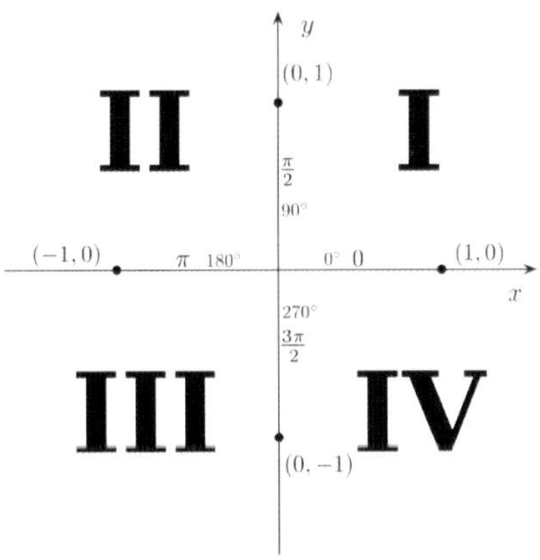

The four quadrants of a Cartesian coordinate system

The table below displays many of the key properties of the sine function (sign, monotonicity, convexity), arranged by the quadrant of the argument. For arguments outside those in the table, one may compute the corresponding information by using the periodicity $\sin(\alpha+360^\circ)=\sin(\alpha)$ of the sine function.

Quadrant	Degrees	Radians	Value	Sign	Monotony	Convexity
1st Quadrant	$0^\circ < x < 90^\circ$	$0 < x < \frac{\pi}{2}$	$0 < \sin(x) < 1$	+	increasing	concave
2nd Quadrant	$90^\circ < x < 180^\circ$	$\frac{\pi}{2} < x < \pi$	$0 < \sin(x) < 1$	+	decreasing	concave
3rd Quadrant	$180^\circ < x < 270^\circ$	$\pi < x < \frac{3\pi}{2}$	$-1 < \sin(x) < 0$	−	decreasing	convex
4th Quadrant	$270^\circ < x < 360^\circ$	$\frac{3\pi}{2} < x < 2\pi$	$-1 < \sin(x) < 0$	−	increasing	convex

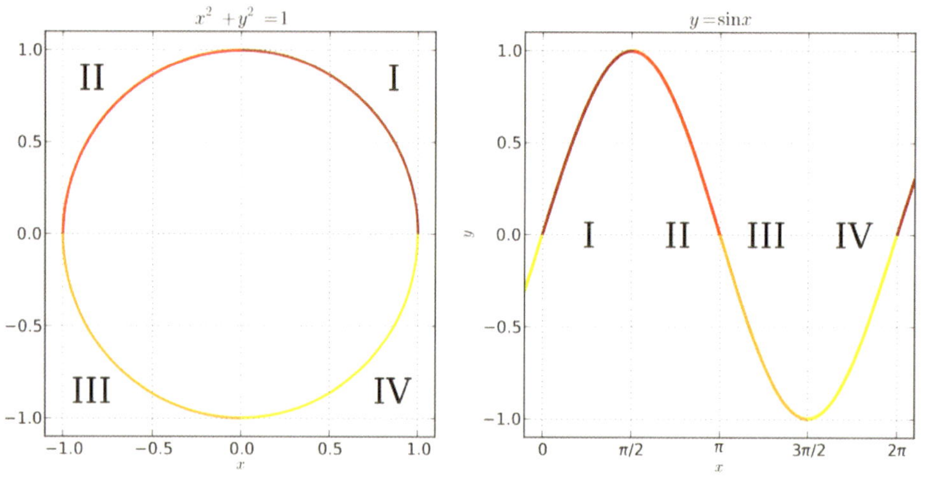

The quadrants of the unit circle and of sin(x), using the Cartesian coordinate system

The following table gives basic information at the boundary of the quadrants.

Degrees	Radians	$\sin(x)$	Point type
0°	0	0	Root, Inflection
90°	$\frac{\pi}{2}$	1	Maximum
180°	π	0	Root, Inflection
270°	$\frac{3\pi}{2}$	−1	Minimum

Fixed Point

Zero is the only real fixed point of the sine function; in other words the only intersection of the sine function and the identity function is $\sin(0) = 0$.

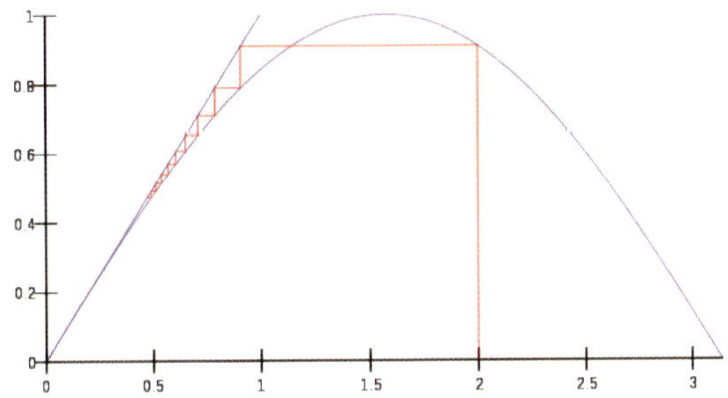

The fixed point iteration $x_{n+1} = \sin(x_n)$ with initial value $x_0 = 2$ converges to 0.

Arc Length

The arc length of the sine curve between a and b is $\int_a^b \sqrt{1+\cos^2(x)}\,dx$ This integral is an elliptic integral of the second kind.

The arc length for a full period is $\frac{4\sqrt{2\pi^3}}{\Gamma(1/4)^2}+\frac{\Gamma(1/4)^2}{\sqrt{2\pi}}=7.640395578\ldots$ where Γ is the Gamma function.

The arc length of the sine curve from 0 to x is the above number divided by 2π times x, plus a correction that varies periodically in x with period π. The Fourier series for this correction can be written in closed form using special functions, but it is perhaps more instructive to write the decimal approximations of the Fourier coefficients. The sine curve arc length from 0 to x is

$$1.21600672x+0.10317093\sin(2x)-0.00220445\sin(4x)+0.00012584$$

$$\sin(6x)-0.00001011\sin(8x)+\cdots$$

Special Values

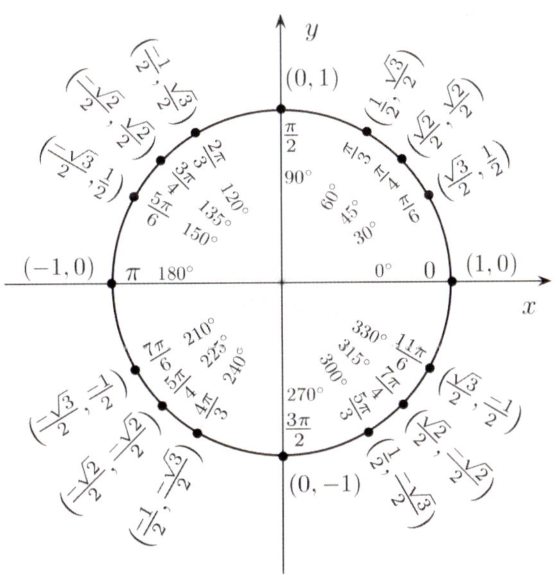

Some common angles (θ) shown on the unit circle. The angles are given in degrees and radians, together with the corresponding intersection point on the unit circle, (cos(θ), sin(θ)).

For certain integral numbers x of degrees, the value of $\sin(x)$ is particularly simple. A table of some of these values is given below.

x (angle)				$\sin(x)$	
Degrees	Radians	Gradians	Turns	Exact	Decimal
0°	0	0^g	0	0	0
180°	π	200^g	$\frac{1}{2}$		
15°	$\frac{1}{12}\pi$	$16\frac{2}{3}^g$	$\frac{1}{24}$	$\frac{\sqrt{6}-\sqrt{2}}{4}$	0.258819045102521
165°	$\frac{11}{12}\pi$	$183\frac{1}{3}^g$	$\frac{11}{24}$		
30°	$\frac{1}{6}\pi$	$33\frac{1}{3}^g$	$\frac{1}{12}$	$\frac{1}{2}$	0.5
150°	$\frac{5}{6}\pi$	$166\frac{2}{3}^g$	$\frac{5}{12}$		
45°	$\frac{1}{4}\pi$	50^g	$\frac{1}{8}$	$\frac{\sqrt{2}}{2}$	0.707106781186548
135°	$\frac{3}{4}\pi$	150^g	$\frac{3}{8}$		
60°	$\frac{1}{3}\pi$	$66\frac{2}{3}^g$	$\frac{1}{6}$	$\frac{\sqrt{3}}{2}$	0.866025403784439
120°	$\frac{2}{3}\pi$	$133\frac{1}{3}^g$	$\frac{1}{3}$		
75°	$\frac{5}{12}\pi$	$83\frac{1}{3}^g$	$\frac{5}{24}$	$\frac{\sqrt{6}+\sqrt{2}}{4}$	0.965925826289068
105°	$\frac{7}{12}\pi$	$116\frac{2}{3}^g$	$\frac{7}{24}$		
90°	$\frac{1}{2}\pi$	100^g	$\frac{1}{4}$	1	1

90 degree increments:

x in degrees	0°	90°	180°	270°	360°
x in radians	0	π/2	π	3π/2	2π
x in gons	0	100^g	200^g	300^g	400^g
x in turns	0	1/4	1/2	3/4	1
$\sin x$	0	1	0	-1	0

Other values not listed above:

$$\sin\left(\frac{\pi}{60}\right) = \sin(3^\circ) = \frac{(2-\sqrt{12})\sqrt{5+\sqrt{5}} + (\sqrt{10}-\sqrt{2})(\sqrt{3}+1)}{16}$$

$$\sin\left(\frac{\pi}{30}\right) = \sin(6^\circ) = \frac{\sqrt{30-\sqrt{180}} - \sqrt{5} - 1}{8}$$

$$\sin\left(\frac{\pi}{20}\right) = \sin(9^\circ) = \frac{\sqrt{10} + \sqrt{2} - \sqrt{20-\sqrt{80}}}{8}$$

$$\sin\left(\frac{\pi}{15}\right) = \sin(12^\circ) = \frac{\sqrt{10+\sqrt{20}} + \sqrt{3} - \sqrt{15}}{8}$$

$$\sin\left(\frac{\pi}{10}\right) = \sin(18^\circ) = \frac{\sqrt{5}-1}{4} = \tfrac{1}{2}\varphi^{-1}$$

$$\sin\left(\frac{7\pi}{60}\right) = \sin(21^\circ) = \frac{(2+\sqrt{12})\sqrt{5-\sqrt{5}} - (\sqrt{10}+\sqrt{2})(\sqrt{3}-1)}{16}$$

$$\sin\left(\frac{}{8}\right) = \sin(22.5^\circ) = \frac{\sqrt{2\quad\sqrt{2}}}{2}$$

$$\sin\left(\frac{2\pi}{15}\right) = \sin(24^\circ) = \frac{\sqrt{3} + \sqrt{15} - \sqrt{10-\sqrt{20}}}{8}$$

$$\sin\left(\frac{3\pi}{20}\right) = \sin(27^\circ) = \frac{\sqrt{20+\sqrt{80}} - \sqrt{10} + \sqrt{2}}{8}$$

$$\sin\left(\frac{11}{60}\right) = \sin(33^\circ) = \frac{(\sqrt{12}-2)\sqrt{5+\sqrt{5}} + (\sqrt{10}-\sqrt{2})(\sqrt{3}+1)}{16}$$

$$\sin\left(\frac{\pi}{5}\right) = \sin(36^\circ) = \frac{\sqrt{10-\sqrt{20}}}{4}$$

$$\sin\left(\frac{13\pi}{60}\right) = \sin(39^\circ) = \frac{(2-\sqrt{12})\sqrt{5-\sqrt{5}} + (\sqrt{10}+\sqrt{2})(\sqrt{3}+1)}{16}$$

$$\sin\left(\frac{7\pi}{30}\right) = \sin(42^\circ) = \frac{\sqrt{30+\sqrt{180}} - \sqrt{5} + 1}{8}$$

Relationship to Complex Numbers

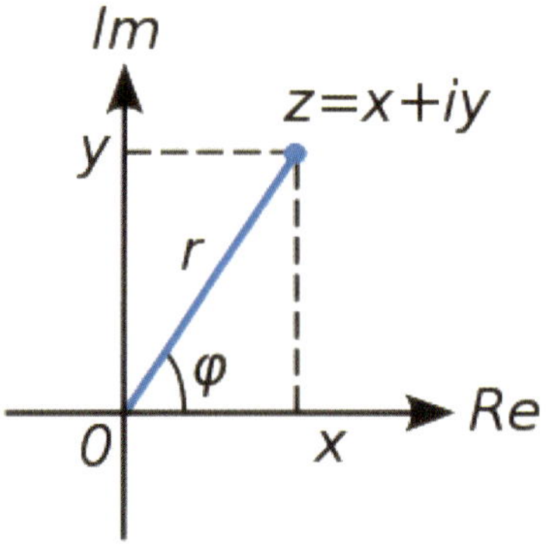

An illustration of the complex plane. The imaginary numbers are on the vertical coordinate axis.

Sine is used to determine the imaginary part of a complex number given in polar coordinates (r, φ):

$$z = r(\cos(\varphi) + i\sin(\varphi))$$

the imaginary part is:

$$\operatorname{Im}(z) = r\sin(\varphi)$$

r and φ represent the magnitude and angle of the complex number respectively. i is the imaginary unit. z is a complex number.

Although dealing with complex numbers, sine's parameter in this usage is still a real number. Sine can also take a complex number as an argument.

Sine with a Complex Argument

$\sin(z)$ Domain coloring of sin(z) over (-π, π) on x- and y-axes. Brightness indicates absolute magnitude, saturation represents complex argument.

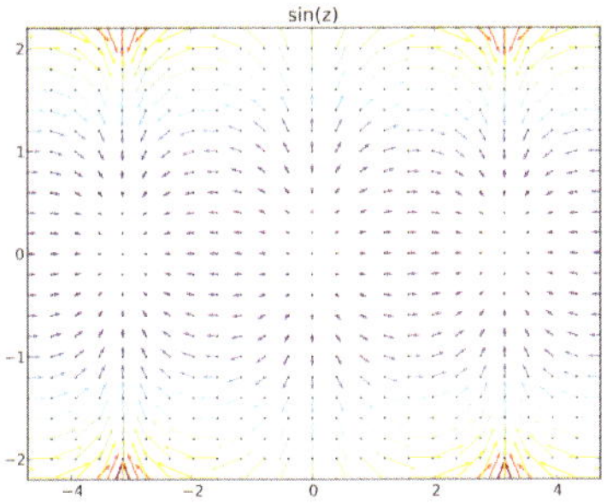

sin(z) as a vector field

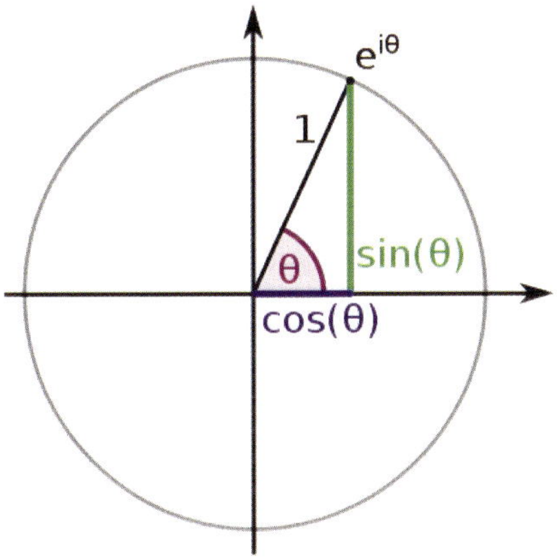

$\sin(\theta)$ is the imaginary part of $e^{i\theta}$.

The definition of the sine function for complex arguments z:

$$\begin{aligned}\sin(z) &= \sum_{n=0}^{\infty} \frac{(-1)^n}{(2n+1)!} z^{2n+1} \\ &= \frac{e^{iz} - e^{-iz}}{2i} \\ &= \frac{\sinh(iz)}{i}\end{aligned}$$

where $i^2 = -1$, and sinh is hyperbolic sine. This is an entire function. Also, for purely real x,

$$\sin(x) = \operatorname{Im}(e^{ix}).$$

For purely imaginary numbers:

$$\sin(iy) = i\sinh(y).$$

It is also sometimes useful to express the complex sine function in terms of the real and imaginary parts of its argument:

$$\begin{aligned}\sin(x + iy) &= \sin(x)\cos(iy) + \cos(x)\sin(iy) \\ &= \sin(x)\cosh(y) + i\cos(x)\sinh(y).\end{aligned}$$

Partial Fraction and Product Expansions of Complex Sine

Using the partial fraction expansion technique in complex analysis, one can find that the infinite series

$$\sum_{n=-\infty}^{\infty}\frac{(-1)^n}{z-n}=\frac{1}{z}-2z\sum_{n=1}^{\infty}\frac{(-1)^n}{n^2-z^2}$$

both converge and are equal to $\frac{\pi}{\sin \pi z}$. Similarly, one can show that

$$\frac{\pi^2}{\sin^2(\pi z)}=\sum_{n=-\infty}^{\infty}\frac{1}{(z-n)^2}.$$

Using product expansion technique, one can derive

$$\sin(\pi z)=\pi z\prod_{n=1}^{\infty}\left(1-\frac{z^2}{n^2}\right).$$

Usage of Complex Sine

sin(*z*) is found in the functional equation for the Gamma function,

$$\Gamma(s)\Gamma(1-s)=\frac{\pi}{\sin(\pi s)},$$

which in turn is found in the functional equation for the Riemann zeta-function,

$$\zeta(s)=2(2\pi)^{s-1}\Gamma(1-s)\sin(\pi s/2)\zeta(1-s).$$

As a holomorphic function, *sin z* is a 2D solution of Laplace's equation:

$$\Delta u(x_1,x_2)=0.$$

It is also related with level curves of pendulum.

Complex Graphs

Sine function in the complex plane		
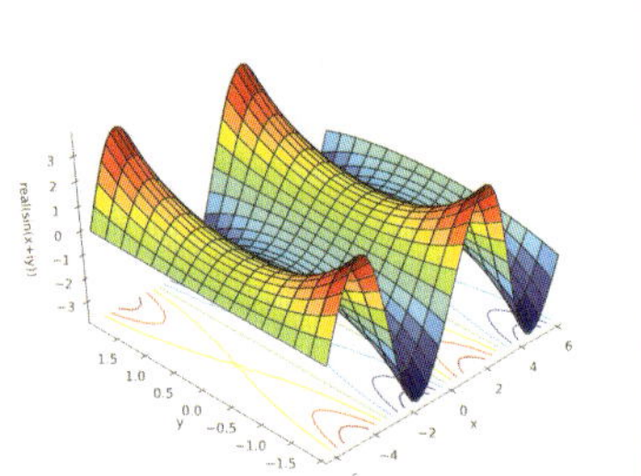	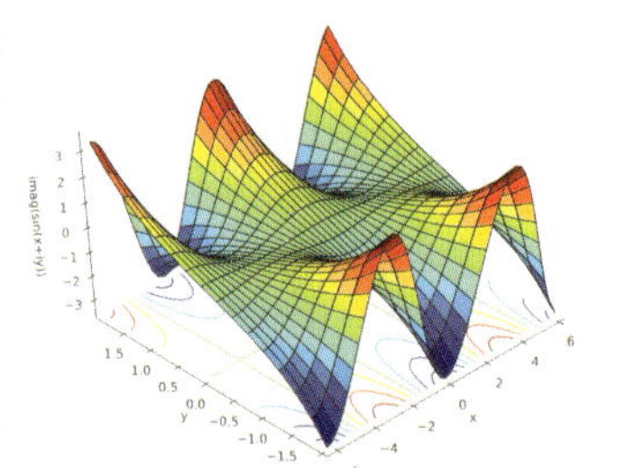	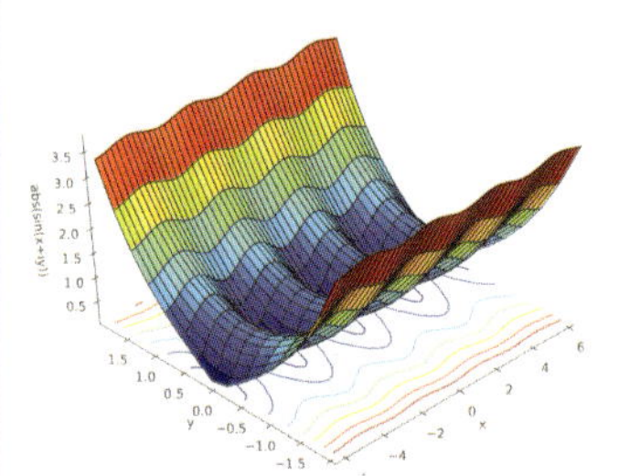
real component	imaginary component	magnitude

Arcsine function in the complex plane		
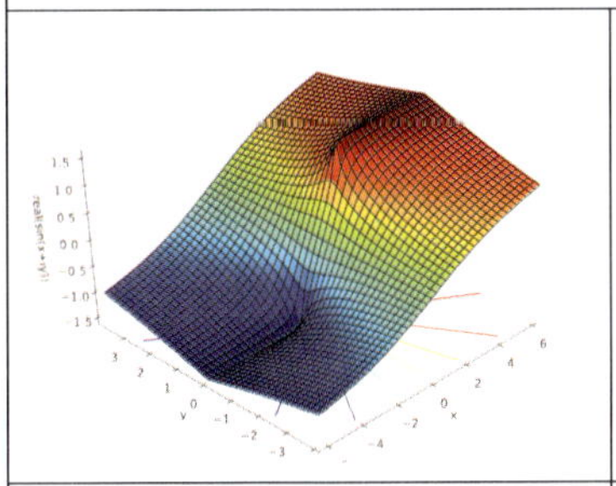	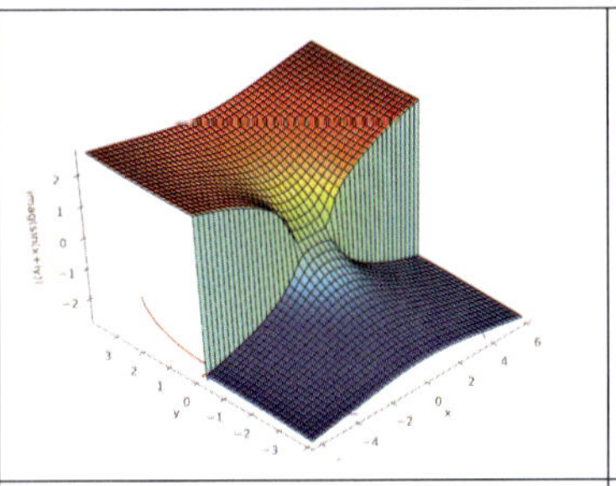	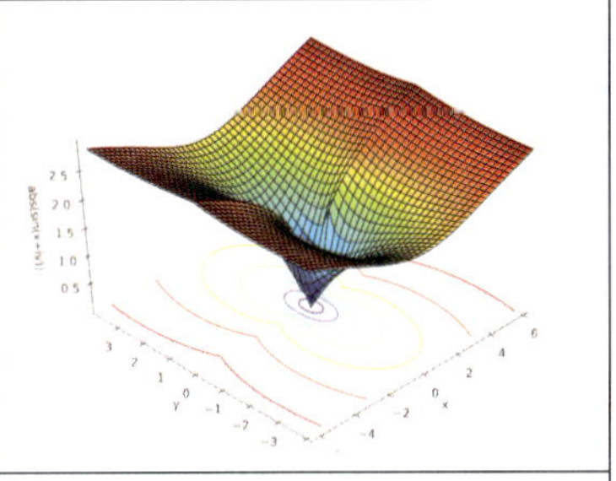
real component	imaginary component	magnitude

Cosine Function

The cosine function is one of the oldest mathematical functions. It was first used in ancient Egypt (c. 2000 B.C.). Much later F. Viète (1590) evaluated some values of $\cos(n\,z)$, E. Gunter introduced the notation "Cosi" and the word "cosinus" (replacing "complementi sinus"), and I. Newton (1658, 1665) found the series expansion for $\cos(z)$.

The classical definition of the cosine function for real arguments is: "the cosine of an angle α in a right-angle triangle is the ratio of the length of the adjacent leg to the length of the hypotenuse." This description of $\cos(\alpha)$ is valid for $0<\alpha<\pi/2$ when the triangle is nondegenerate. This approach to the cosine can be expanded to arbitrary real values of α if consideration is given to the arbitrary point (x,y) in the x,y-Cartesian plane and $\cos(\alpha)$ is defined as the ratio $x/(x^2,y^2)^{1/2}$, assuming that α is the value of the angle between the positive direction of the x-axis and the direction from the origin to the point (x,y).

The following formula can also be used as a definition of the cosine function:

$$\cos(z)=1-\frac{z^2}{2}+\frac{z^4}{24}-\ldots=\sum_{k=0}^{\infty}\frac{(-1)^k z^{2k}}{(2k)!}.$$

This series converges for all finite numbers z.

A Quick Look at the Cosine Function

Here is a graphic of the cosine function $f(x)=\cos(x)$ for real values of its argument x.

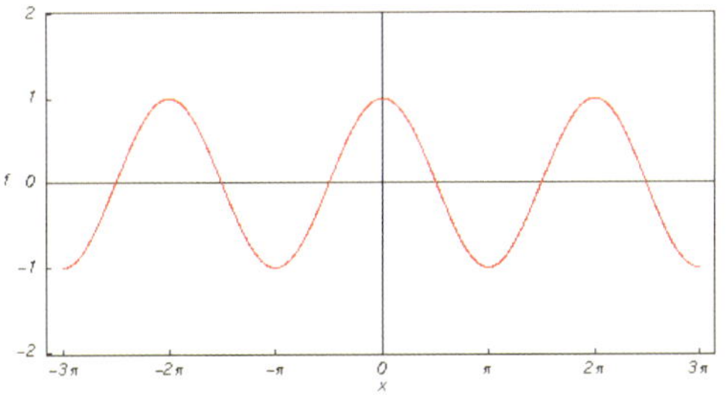

Representation Through More General Functions

The function $\cos(z)$ is a particular case of more complicated mathematical functions. For example, it is a special case of the generalized hyper geometric function ${}_0F_1(;a;w)$ with the parameter $a=\frac{1}{2}$ at $w=-\frac{z^2}{4}$:

$$\cos(z)={}_0F_1\left(;\frac{1}{2};-\frac{z^2}{4}\right).$$

It is also a particular case of the Bessel function $J_Y(z)$ with the parameter $Y=-\frac{1}{2}$, multiplied by $\sqrt{\frac{\pi z}{2}}$:

$$\cos(z)=\sqrt{\frac{\pi z}{2}}J_{-\frac{1}{2}}(z).$$

Other Bessel functions can also be expressed through cosine functions for similar values of the parameter:

$$\cos(z)=\sqrt{\frac{\pi iz}{2}}I_{-\frac{1}{2}}(iz)\ \cos(z)=-\sqrt{\frac{\pi z}{2}}\ Y_{\frac{1}{2}}(z)\ \cos(z)=\frac{1}{\sqrt{2\pi}}\left(\sqrt{iz}\ K_{\frac{1}{2}}(iz)+\sqrt{-iz}K_{\frac{1}{2}}(-iz)\right).$$

Struve functions can also degenerate into the cosine function for similar values of the parameter:

$$\cos(z)=1-\sqrt{\frac{\pi z}{2}}H_{\frac{1}{2}}(z)\ \cos(z)=1+\sqrt{\frac{\pi iz}{2}}L_{\frac{1}{2}}(iz).$$

But the function $\cos(z)$ is also a degenerate case of the doubly periodic Jacobi elliptic functions when their second parameter is equal to 0 or :

$$\cos(z)=\operatorname{cd}(z|0)=\operatorname{cn}(z|0)$$
$$\cos(z)=\operatorname{nc}(i\,z|1)=\operatorname{nd}(i\,z|1).$$

Finally, the function $\cos(z)$ is the particular case of another class of functions—the Mathieu functions:

$$\cos(z)=\operatorname{ce}(1,0,z).$$

Definition of the Cosine Function for a Complex Argument

In the complex z-plane, the function $\cos(z)$ is defined using the exponential function e^w in the points $w=iz$ and $w=-iz$ through the formula:

$$\cos(z)=\frac{e^{iz}+e^{-iz}}{2}.$$

The key role in this definition of $\cos(z)$ belongs to the famous Euler formula that connects the exponential, the sine, and the cosine functions:

$$e^{iz} = \cos(z) + i\sin(z).$$

Changing z to $-z$, the Euler formula can be converted into the following modification:

$$e^{-iz} = \cos(z) - i\sin(z).$$

Adding the preceding formulas gives the following result:

$$\cos(z) = \frac{e^{iz} + e^{-iz}}{2}.$$

Here are two graphics showing thc real and imaginary parts of the cosine function over the complex plane.

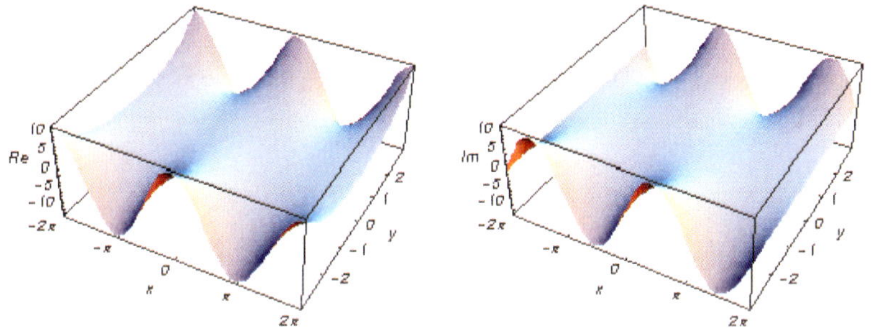

The Best-known Properties and Formulas for the Cosine Function

Values in Points

Students usually learn the following basic table of cosine function values for special points of the circle: $\cos(0)=1$ $\cos\left(\frac{\pi}{6}\right)=\frac{\sqrt{3}}{2}$ $\cos\left(\frac{\pi}{4}\right)=\frac{1}{\sqrt{2}}$ $\cos\left(\frac{\pi}{3}\right)=\frac{1}{2}$

$\cos\left(\frac{\pi}{2}\right)=0$ $\cos\left(\frac{2\pi}{3}\right)=-\frac{1}{2}$ $\cos\left(\frac{3\pi}{4}\right)=-\frac{1}{\sqrt{2}}$ $\cos\left(\frac{5\pi}{6}\right)=-\frac{\sqrt{3}}{2}$

$\cos(\pi)=-1$ $\cos\left(\frac{7\pi}{6}\right)=-\frac{\sqrt{3}}{2}$ $\cos\left(\frac{5\pi}{4}\right)=-\frac{1}{\sqrt{2}}$ $\cos\left(\frac{4\pi}{3}\right)=-\frac{1}{2}$

$\cos\left(\frac{3\pi}{2}\right)=0$ $\cos\left(\frac{5\pi}{3}\right)=\frac{1}{2}$ $\cos\left(\frac{7\pi}{4}\right)=\frac{1}{\sqrt{2}}$ $\cos\left(\frac{11\pi}{6}\right)=\frac{\sqrt{3}}{2}$

$\cos(2\pi)=1$ $\cos(\pi m)=(-1)^m /; m\in\mathbb{Z}$ $\cos\left(\pi\left(\frac{1}{2}+m\right)\right)=0/; m\in\mathbb{Z}.$

General Characteristics

For real values of argument z, the values of $\cos(z)$ are real.

In the points $z = 2\pi n/m /; n \in \mathbb{Z} \wedge m \in \mathbb{Z}$, the values of $\cos(z)$ are algebraic. In several cases they can even be rational numbers, 0, or 1. Here are some examples:

$$\cos(0) = 1 \quad \cos\left(\frac{\pi}{3}\right) = \frac{1}{2} \quad \cos\left(\frac{\pi}{2}\right) = 0.$$

The values of $\cos\left(\frac{n\pi}{m}\right)$ can be expressed using only square roots if $n \in \mathbb{Z}$ and m is a product of a power of 2 and distinct Fermat primes {3, 5, 17, 257, ...}.

The function $\cos(z)$ is an entire analytical function of z that is defined over the whole complex z -plane and does not have branch cuts and branch points. It has an essential singular point at $z = \tilde{\infty}$. It is a periodic function with the real period 2π:

$$\cos(z + 2\pi) = \cos(z)$$

$$\cos(z) = \cos(z + 2\pi k) /; k \in \mathbb{Z} \quad \cos(z) = (-1)^k \cos(z + \pi k) /; k \in \mathbb{Z}.$$

The function $\cos(z)$ is an even function with mirror symmetry:

$$\cos(-z) = \cos(z) \quad \cos(\bar{z}) = \overline{\cos(z)}.$$

Differentiation

The derivatives of $\cos(z)$ have simple representations using either the $\sin(z)$ function or the $\cos(z)$ function:

$$\frac{\partial \cos(z)}{\partial z} = -\sin(z) \quad \frac{\partial^n \cos(z)}{\partial z^n} = \cos\left(z + \frac{\pi n}{2}\right) /; n \in \mathbb{N}^+.$$

Ordinary Differential Equation

The function cos( ) satisfies the simplest possible linear differential equation with constant coefficients:

$$w''(z) + w(z) = 0 /; w(z) = \cos(z) \wedge w(0) = 1 \wedge w'(0) = 0.$$

The complete solution of this equation can be represented as a linear combination of $\sin(z)$ and $\cos(z)$ with arbitrary constant coefficients c_1 and c_2:

$$w''(z) + w(z) = 0 /; w(z) = c_1 \cos(z) + c_2 \sin(z).$$

The function $\cos(z)$ also satisfies first-order nonlinear differential equations:

$$w'(z)-\sqrt{1-(w(z))^2}=0/;w(z)=\cos(z)\wedge w(0)=1\wedge \operatorname{Re}(z)<\frac{\pi}{2}.$$

Series Representation

The function $\cos(z)$ has a simple series expansion at the origin that converges in the whole complex z-plane:

$$\cos(z)=1-\frac{z^2}{2}+\frac{z^4}{24}-\ldots=\sum_{k=0}^{\infty}\frac{(-1)^k z^{2k}}{(2k)!}.$$

For real $z=x$ this series can be interpreted as the real part of the series expansion for the exponential function e^{ix}:

$$\cos(z)=1-\frac{z^2}{2}+\frac{z^4}{24}-\ldots=\operatorname{Re}\left(1+ix+\frac{(ix)^2}{2!}+\frac{(ix)^3}{3!}+\ldots\right)=\operatorname{Re}\left(\sum_{k=0}^{\infty}\frac{(ix)^k}{k!}\right)=\operatorname{Re}\left(e^{ix}\right).$$

Product Representation

The following famous infinite product representation for $\cos(z)$ clearly illustrates that $\cos(z)=0$ at $z=\pi k-\frac{\pi}{2}\wedge k\in\mathbb{Z}$:

$$\cos(z)=\prod_{k=1}^{\infty}\left(1-\frac{4z^2}{\pi^2(2k-1)^2}\right)$$

Indefinite Integration

Indefinite integrals of expressions involving the cosine function can sometimes be expressed using elementary functions. However, special functions are frequently needed to express the results even when the integrands have a simple form (if they can be evaluated in closed form). Here are some examples:

$$\int\cos(z)dz=\sin(z)$$

$$\int\sqrt{\cos(z)}\,dz-2E\left(\frac{z}{2}\middle|2\right)$$

$$\int z^{\alpha-1}\cos^{\nu}(az)dz=\frac{2^{-\nu}(1-\nu \bmod 2)}{\alpha}z^{\alpha}\binom{\nu}{\frac{\nu}{2}}-$$

$$2^{-\nu}z^{a}\sum_{j=0}^{\left\lceil\frac{\nu-1}{2}\right\rceil}\binom{\nu}{j}\left(\Gamma(\alpha,ia(2j-\nu)z)(ia(2j-\nu)z)^{-a}+\Gamma(\alpha,ia(\nu-2j)z)(ia(\nu-2j)z)^{-a}\right)/;\alpha\neq0\wedge\nu\in\mathbb{N}^{+}.$$

The last integral cannot be evaluated in closed form using the known classical special functions for arbitrary values of parameters α and ν.

Definite Integration

Definite integrals that contain the cosine function are sometimes simple. For example, the famous Dirichlet type and Fresnel integrals have the following values:

$$\int_0^\infty \frac{\cos(t)-1}{t}\,dt=-\gamma$$

$$\int_0^\infty \cos(t^2)\,dt=\frac{1}{2}\sqrt{\frac{\pi}{2}},$$

Where γ is the Euler-Mascheroni constant $\gamma=0.577216\ldots$.

Some special functions can be used to evaluate more complicated definite integrals. For example, elliptic integrals and gamma functions are needed to express the following integrals:

$$\int_0^\pi \cos^{\frac{1}{2}}(z)\,dz=2\,E(\,2)$$

$$\int_0^\pi z^{\alpha-1}\cos(z)\,dz=\cos\left(\frac{\pi\alpha}{2}\right)\Gamma(\alpha)/;0<\mathrm{Re}(\alpha)<1.$$

Integral Transforms

Integral transforms of expressions involving the cosine function may not be classically convergent but can be interpreted in a generalized functions (distributions) sense. For example, the exponential Fourier transform of the cosine function $\cos(z)$ does not exist in the classical sense but can be expressed using the Dirac delta function.

$$\mathcal{F}_t[\cos(t)](z)=\sqrt{\frac{\pi}{2}}\,\delta(z-1)+\sqrt{\frac{\pi}{2}}\,\delta(z+1).$$

Among other integral transforms of the cosine function, the best known are the Fourier cosine and sine transforms, and the Laplace, Mellin, Hilbert, and Hankel transforms:

$$\mathcal{F}c_t\left[\cos(a\,t)\right](z)=\sqrt{\frac{\pi}{2}}\left(\delta(a-z)+\delta(a+z)\right)/;a\in\mathbb{R}$$

$$\mathcal{F}s_t\left[\cos(a\,t)\right](z)=-\sqrt{\frac{2}{\pi}}\frac{z}{a^2-z^2}/;a\in\mathbb{R}$$

$$\mathcal{L}_t\left[\cos(t)\right](z)=\frac{z}{1+z^2}$$

$$\mathcal{M}_t\left[\cos(t)\right](z)=\Gamma(z)\cos\left(\frac{\pi z}{2}\right)/;0<\mathrm{Re}(z)<1$$

$$\mathcal{H}_t\left[\cos(t)\right](x)=-\sin(x)$$

$$\mathcal{H}_{t;y}\left[t^{\alpha-1}\cos(t)\right](z)=$$

$$\frac{1}{\Gamma(\nu+1)}\left(2^{-\nu} z^{\nu+\frac{1}{2}} \cos\left(\frac{1}{4}\pi(2\alpha+2\nu+1)\right)\Gamma\left(\alpha+\nu+\frac{1}{2}\right){}_2F_1\left(\frac{1}{4}(2\alpha+2\nu+1),\frac{1}{4}(2\alpha+2\nu+3);\nu+1;z^2\right)\right)/;$$

$$\mathrm{Re}(\alpha+\nu)>-\frac{1}{2}\wedge\mathrm{Re}(\alpha)<1.$$

Finite Summation

The following finite sums from the cosine can be expressed using the trigonometric functions:

$$\sum_{k=0}^{n}\cos(a\,k)=\csc\left(\frac{a}{2}\right)\sin\left(\frac{a(n+1)}{2}\right)\cos\left(\frac{an}{2}\right)$$

$$\sum_{k=0}^{n}(-1)^k\cos(a\,k)=\cos\left(\frac{1}{2}(a+n(a+\pi))\right)\sec\left(\frac{a}{2}\right)\cos\left(\frac{1}{2}n(a+\pi)\right)$$

$$\sum_{k=0}^{n}\cos(a\,k+z)=\csc\left(\frac{a}{2}\right)\sin\left(\frac{1}{2}a(n+1)\right)\cos\left(\frac{an}{2}+z\right)$$

$$\sum_{k=0}^{n}(-1)^k\cos(a\,k+z)=\cos\left(\frac{1}{2}(a+n(a+\pi))\right)\sec\left(\frac{a}{2}\right)\cos\left(\frac{1}{2}n(a+\pi)+z\right)$$

$$\sum_{k=1}^{n}z^k\cos(\,k\,a)=\frac{z\left(\cos(a\,n)z^{n+1}-\cos(n\,a+a)z^n-z+\cos(a)\right)}{z^2-2\cos(a)z+1}.$$

Infinite Summation

The following infinite sums can be expressed using elementary functions:

$$\sum_{k=1}^{\infty}\frac{\cos(a\,k)}{k}=\frac{1}{2}\log\left(\frac{1}{2(1-\cos(a))}\right)/;0<\mathrm{Re}(a)<2\pi$$

$$\sum_{k=1}^{\infty}\frac{(-1)^{k-1}\cos(a\,k)}{k}=\log\left(2\cos\left(\frac{a}{2}\right)\right)/;|\mathrm{Re}(a)|<\pi$$

$$\sum_{k=0}^{\infty}\frac{\cos(a\,k)}{k!}=e^{\cos(a)}\cos(\sin(a))$$

$$\sum_{k=0}^{\infty}\frac{z^k\cos(a\,k)}{k!}=e^{z\cos(a)}\cos(z\sin(a))$$

$$\sum_{k=0}^{\infty}z^k\cos(a\,k)=\frac{1-z\cos(a)}{z^2-2\cos(a)z+1}/;|z|<1.$$

Finite Products

The following finite products from the cosine can be expressed using trigonometric functions:

$$\prod_{k=1}^{2n-1}\cos\left(\frac{k\pi}{n}\right)=2^{1-2n}\left((-1)^n-1\right)/;n\in\mathbb{N}^+$$

$$\prod_{k=1}^{\left\lfloor \frac{n-1}{2} \right\rfloor} \cos\left(\frac{\pi k}{n}\right) = 2^{\frac{1-n}{2}} \left(\frac{1}{2}\left(1-(-1)^n\right) + \frac{1}{2}\left(1+(-1)^n\right)\sqrt{n}\right) /; n \in \mathbb{N}^+$$

$$\prod_{k=1}^{n-1} \cos\left(\frac{\pi k}{n} + z\right) = \frac{2^{1-n}}{\cos(z)} \sin\left(n\left(z + \frac{\pi}{2}\right)\right) /; n \in \mathbb{N}^+$$

$$\prod_{k=1}^{n-1} \cos\left(\frac{2\pi k}{n} + z\right) = (-2)^{1-n} \sec(z)\left(\cos(n\,z) - \cos\left(\frac{n\pi}{2}\right)\right) /; n \in \mathbb{N}^+.$$

Infinite Products

The following infinite product that contains the cosine function can be expressed using the sine function:

$$\prod_{k=1}^{\infty} \cos\left(\frac{z}{2^k}\right) = \frac{\sin(z)}{z}.$$

Addition Formulas

The cosine of a sum can be represented by the rule: “the cosine of a sum is equal to the product of the cosines minus the product of the sines.” A similar rule is valid for the cosine of the difference:

$$cos(a+b) = cos(a)cos(b) - sin(a)sin(b)$$
$$cos(a-b) = cos(a)cos(b) + sin(a)sin(b).$$

Multiple Arguments

In the case of multiple arguments z, $2z$, $3z$, ..., the function $\cos(z)$ can be represented as the finite sum of terms that include powers of the sine and the cosine:

$$\cos(2z) = \cos^2(z) - \sin^2(z)$$

$$\cos(3z) = 4\cos^3(z) - 3\cos(z)$$

$$\cos(n\,z) = \sum_{k=0}^{\left\lfloor \frac{n}{2} \right\rfloor} (-1)^k \binom{n}{2k} \sin^{2k}(z) \cos^{n-2k}(z) /; n \in \mathbb{N}^+.$$

The function $\cos(n\,z)$ can also be represented as the finite sum including only the cosine of z:

$$\cos(n\,z) = n \sum_{k=0}^{\left\lfloor \frac{n}{2} \right\rfloor} \frac{(-1)^k (n-k-1)!\, 2^{n-2k-1}}{k!(n-2k)!} \cos^{n-2k}(z) /; n \in \mathbb{N}^+.$$

Half-angle Formulas

The cosine of the half-angle can be represented by the following simple formula that is valid in some vertical strips:

$$\cos\left(\frac{z}{2}\right)=\sqrt{\frac{1+\cos(z)}{2}}\,/;\,|\mathrm{Re}(z)|<\pi\vee\mathrm{Re}(z)=-\pi\wedge\mathrm{Im}(z)\geq 0\vee\mathrm{Re}(z)=\pi\wedge\mathrm{Im}(z)\leq 0.$$

To make this formula correct for all complexes z, a complicated prefactor is needed:

$$\cos\left(\frac{z}{2}\right)=c(z)\sqrt{\frac{1+\cos(z)}{2}}\,/;\,c(z)=(-1)^{\left\lfloor\frac{\mathrm{Re}(z)+\pi}{2\pi}\right\rfloor}\left(1-\left(1+(-1)^{\left\lfloor\frac{\mathrm{Re}(z)+\pi}{2\pi}\right\rfloor+\left\lfloor-\frac{\mathrm{Re}(z)+\pi}{2\pi}\right\rfloor}\right)\theta(-\mathrm{Im}(z))\right),$$

Where $c(z)$ contains the unit step, real part, imaginary part, and the floor functions.

Sums of two Direct Functions

The sum of two cosine functions can be described by the rule: "the sum of the cosines is equal to two times the cosine of the half-difference multiplied by the cosine of the half-sum." A similar rule is valid for the difference of two cosines:

$$\cos(a)+\cos(b)=2\cos\left(\frac{a-b}{2}\right)\cos\left(\frac{a+b}{2}\right)$$

$$\cos(a)-\cos(b)=-2\sin\left(\frac{a+b}{2}\right)\sin\left(\frac{a-b}{2}\right).$$

Products Involving the Direct Function

The product of two cosine functions and the product of the cosine and sine have the following representations:

$$\cos(a)\cos(b)=\frac{1}{2}(\cos(a-b)+\cos(a+b))$$

$$\cos(a)\sin(b)=\frac{1}{2}(\sin(b-a)+\sin(a+b)).$$

Powers of the Direct Function

The integer powers of the cosine functions can be expanded as finite sums of cosine functions with multiple arguments. These sums include binomial coefficients:

$$\cos^n(z)=2^{-n}\binom{n}{\frac{n}{2}}(1-n \bmod 2)+2^{1-n}\sum_{k=0}^{\left\lfloor\frac{n-1}{2}\right\rfloor}\binom{n}{k}\cos((n-2k)z)\,/;\,n\in\mathbb{N}^{+}.$$

Inequalities

The best-known inequalities for cosine functions are the following:

$$|cos(x)| \leq 1/; x \in \mathbb{R}$$

$$cos(x) < \frac{sin(x)}{x} /; 0 < x < 4.493409.$$

Relations with its Inverse Function

There are simple relations between the function $\cos(z)$ and its inverse function $\cos^{-1}(z)$:

$$\cos^{-1}(\cos(z)) = z \ \cos^{-1}(\cos(z)) = z /; 0 < \mathrm{Re}(z) < \pi \vee \mathrm{Re}(z) = 0 \wedge \mathrm{Im}(z) \geq 0 \vee \mathrm{Re}(z) = \pi \wedge \mathrm{Im}(z) \leq 0.$$

The second formula is valid at least in the vertical strip $0 < \mathrm{Re}(z) < \pi$. Outside of this strip, a much more complicated relation (that contains the unit step, real part, and the floor functions) holds:

$$\cos^{-1}(\cos(z)) = \frac{\pi}{2}\left(1 - (-1)^{\left\lfloor -\frac{\mathrm{Re}(z)}{\pi} \right\rfloor}\right) + (-1)^{\left\lfloor -\frac{\mathrm{Re}(z)}{\pi} \right\rfloor}\left(\left(1 + (-1)^{\left\lfloor \frac{\mathrm{Re}(z)}{\pi} \right\rfloor + \left\lfloor -\frac{\mathrm{Re}(z)}{\pi} \right\rfloor}\right)\theta(\mathrm{Im}(z)-1)\left(z + \pi\left\lfloor -\frac{\mathrm{Re}(z)}{\pi} \right\rfloor\right)\right).$$

Representations Through other Trigonometric Functions

Cosine and sine functions are connected by a very simple formula including the linear function in the argument:

$$\cos(z) = \sin\left(\frac{\pi}{2} - z\right).$$

Another famous formula, connecting $\cos(z)$ and $\sin(z)$, is shown in the well-known Pythagorean theorem:

$$\cos^2(z) = 1 - \sin^2(z)$$

$$\cos(z) = \sqrt{1 - \sin^2(z)} /; |\mathrm{Re}(z)| < \frac{\pi}{2} \bigvee \mathrm{Re}(z) = -\frac{\pi}{2} \bigwedge \mathrm{Im}(z) \geq 0 \bigvee \mathrm{Re}(z) = \frac{\pi}{2} \bigwedge \mathrm{Im}(z) \leq 0.$$

The last restriction on z can be removed, but the formula will get a complicated coefficient $c(z)$ with $|c(z)| = 1$, that contains the unit step, real part, imaginary part, and the floor function:

$$\cos(z) = c(z)\sqrt{1 - \sin^2(z)} /; c(z) = (-1)^{\left\lfloor \frac{1}{2} - \frac{\mathrm{Re}(z)}{\pi} \right\rfloor}\left(1 - \left(1 + (-1)^{\left\lfloor \frac{\mathrm{Re}(z)}{\pi} - \frac{1}{2} \right\rfloor + \left\lfloor \frac{1}{2} - \frac{\mathrm{Re}(z)}{\pi} \right\rfloor}\right)\theta(\mathrm{Im}(z))\right).$$

The cosine function can also be represented using other trigonometric functions by the following formulas:

$$\cos(z)=\frac{1-\tan^2\left(\frac{z}{2}\right)}{1+\tan^2\left(\frac{z}{2}\right)} \quad \cos(z)=\frac{\cot^2\left(\frac{z}{2}\right)-1}{\cot^2\left(\frac{z}{2}\right)+1}$$

$$\cos(z)=\frac{1}{\csc\left(\frac{\pi}{2}-z\right)} \quad \cos(z)=\frac{1}{\sec(z)}.$$

Representations Through Hyperbolic Functions

The cosine function has representations using the hyperbolic functions:

$$\cos(z)=\cosh(iz) \quad \cos(iz)=\cosh(z) \; \cos(z)=-i\sinh\left(\frac{\pi i}{2}-iz\right)$$

$$\cos(z)=\frac{1+\tanh^2\left(\frac{zi}{2}\right)}{1-\tanh^2\left(\frac{zi}{2}\right)} \quad \cos(z)=\frac{\coth^2\left(\frac{iz}{2}\right)+1}{\coth^2\left(\frac{iz}{2}\right)-1} \quad \cos(z)=-\frac{i}{\operatorname{csch}\left(\frac{\pi i}{2}-iz\right)} \quad \cos(z)=\frac{1}{\operatorname{sech}(iz)}.$$

Tangent Function

The tangent function is a periodic function which is very important in trigonometry.

The tangent function, along with sine and cosine, is one of the three most common trigonometric functions. In any right triangle, the tangent of an angle is the length of the opposite side (O) divided by the length of the adjacent side (A). In a formula, it is written simply as 'tan'.

Often remembered as "TOA" - meaning Tangent is Opposite over Adjacent.

$$\tan x=\frac{O}{A}$$

As an example, let's say we want to find the tangent of angle C in the figure above (click 'reset' first). From the formula above we know that the tangent of an angle is the opposite side divided by the adjacent side. The opposite side is AB and has a length of 15. The adjacent side is BC with a length of 26. So we can write

$$\tan C=\frac{15}{26}$$

This division on the calculator comes out to 0.577. So we can say "*The tangent of C is 0.5776* " or *tan C=0.577*

The simplest way to understand the tangent function is to use the unit circle. For a given angle measure θ draw a unit circle on the coordinate plane and draw the angle centered at the origin, with one side as the positive x-axis. The x-coordinate of the point where the other side of the angle intersects the circle is cos(θ) and the y-coordinate is sin(θ).

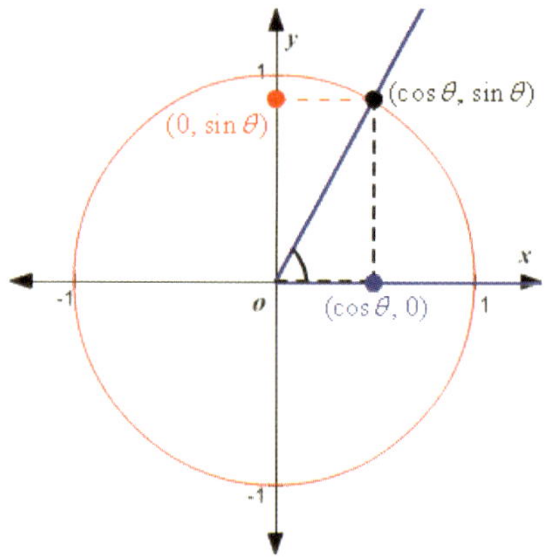

There are a few sine and cosine values that should be memorized, based on 30°–60°–90°triangles and 45°–45°–90°triangles. Based on these, you can work out the related values for tangent.

sin(θ)	cos(θ)	tan(θ)
$\sin(0^\circ)=0$	$\cos(0^\circ)=1$	$\tan(0^\circ)=-=0$
$\sin(30^\circ)=\frac{1}{2}$	$\cos(30^\circ)=\frac{\sqrt{3}}{2}$	$\tan(30^\circ)=\frac{1}{2}\cdot\frac{2}{\sqrt{3}}=\frac{\sqrt{3}}{3}$
$\sin(45^\circ)=\frac{\sqrt{2}}{2}$	$\cos(45^\circ)=\frac{\sqrt{2}}{2}$	$\tan(45^\circ)=\frac{\sqrt{2}}{2}\cdot\frac{2}{\sqrt{2}}=1$
$\sin(60^\circ)=\frac{\sqrt{3}}{2}$	$\cos(60^\circ)=\frac{1}{2}$	$\tan(60^\circ)=\frac{\sqrt{3}}{2}\cdot\frac{2}{1}=\sqrt{3}$
$\sin(90^\circ)=1$	$\cos(90^\circ)=0$	$\tan(90^\circ)=\frac{1}{0}=\text{undef.}$

Note that:

- for angles with their terminal arm in Quadrant II, since sine is positive and cosine is negative, tangent is negative.
- for angles with their terminal arm in Quadrant III, since sine is negative and cosine is negative, tangent is positive.
- for angles with their terminal arm in Quadrant IV, since sine is negative and cosine is positive, tangent is negative.

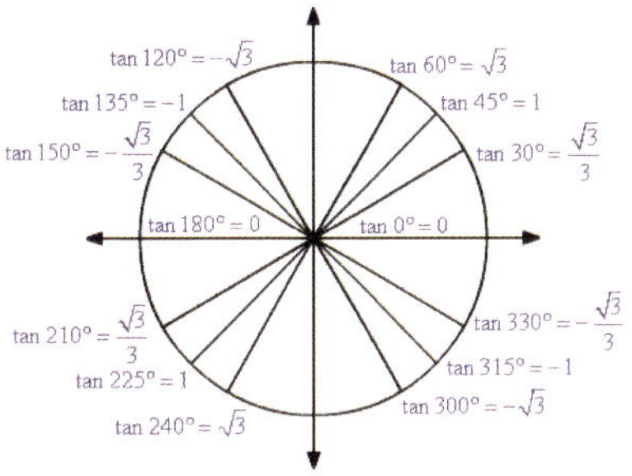

You can plot these points on a coordinate plane to show part of the function, the part between 0 and 2π.

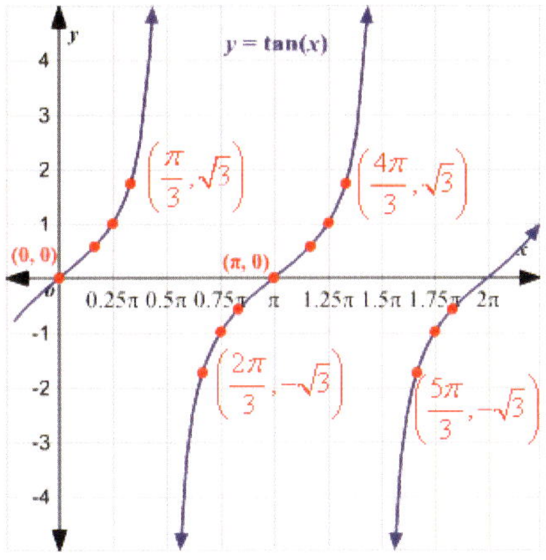

For values of θ less than 0 or greater than 2π you can find the value of θ using the reference angle.

The graph of the function over a wider interval is shown below.

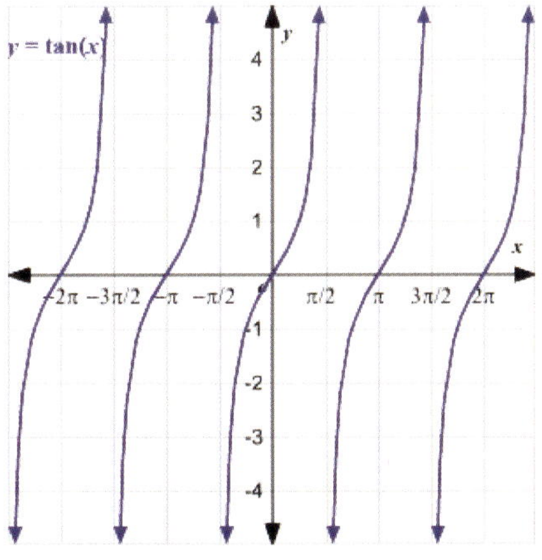

Note that the domain of the function is the whole real line and the range is $-\infty \leq y \leq \infty$.

Example - using Tangent to find a Side Length

If we look at the general definition –

$$\tan x = \frac{O}{A}$$

We see that there are three variables: the measure of the angle *x*, and the lengths of the two sides (Opposite and Adjacent). So if we have any two of them, we can find the third.

Imagine we didn't know the length of the side BC. We know that the tangent of A (60°) is the opposite side divided by the adjacent side AB - the one we are trying to find.

$$\tan 60° = \frac{BC}{15}$$

From our calculator we find that tan 60° is 1.733, so we can write

$$1.733 = \frac{BC}{15}$$

Transposing:

$BC = 15 \times 1.733$

Which comes out to 26, which matches the figure above?

The Inverse Tangent Function - arctan

For every trigonometry function such as tan, there is an inverse function that works in reverse. These inverse functions have the same name but with 'arc' in front. So the inverse of tan is arctan etc.

When we see "arctan A", we interpret it as "the angle whose tangent is A"

tan 60 = 1.733	Means: The tangent of 60 degrees is 1.733
arctan 1.733 = 60	Means: The angle whose tangent is 1.733 is 60 degrees.

We use it when we know what the tangent of an angle is, and want to know the actual angle.

Large and Negative Angles

In a right triangle, the two variable angles are always less than 90° .But we can in fact find the tangent of any angle, no matter how large, and also the tangent of negative angles.

The Derivative of tan(x)

In calculus, the derivative of $tan(x)$ is $sec^2(x)$. This means that at any value of *x*, the rate of change or slope of $tan(x)$ is $sec^2(x)$.

Cotangent Function

The cotangent function is a trigonometric function which is one of three reciprocal functions.

A reciprocal function is one that is the *reciprocal* (or *multiplicative inverse*) of another function (see below). The *cotangent function* (usually abbreviated as cot) is the reciprocal function of the *tangent function*. As with the other trigonometric functions in modern use, the cotangent function was certainly known to Arab scholars, and tables of cotangent values are known to have been in existence by the tenth century CE. In order to clarify exactly how the cotangent function and the tangent function are related, consider the diagram below.

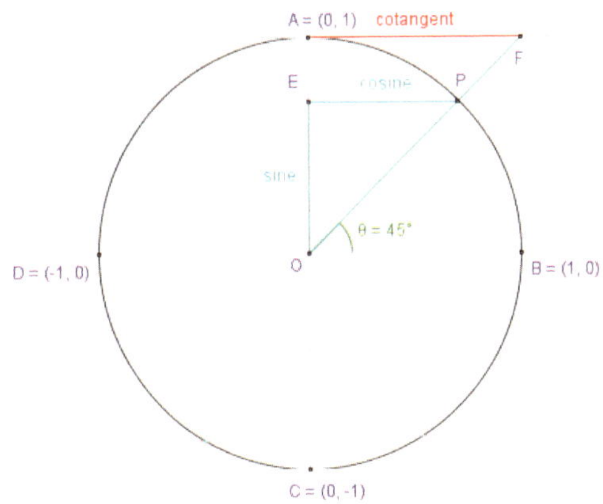

The cotangent as a line segment of the unit circle

Like the other trigonometric functions, the cotangent can be represented as a line segment associated with the *unit circle*. The diagram shows the cotangent for an angle of rotation θ of *forty-five degrees* (measured anti-clockwise from the positive *x*-axis). Line segment *AF* (shown in red) is the cotangent, and lies on the line that is tangent to the circle at point *A*. Line segment *PF* is an extension of line segment *OP* (and, incidentally, also happens to be the *secant*). Remember that point P is a point on the circumference of the unit circle whose *x* and *y* coordinates represent the value of cos (θ) and sin (θ) respectively (the line segments representing the sine and cosine are also shown). Remember that one definition of the tangent function is as the *quotient* of the *sine and cosine functions*.

Remember that the tangent function can be defined either as the quotient of the opposite and adjacent, or as the quotient of the sine and cosine functions the value of sin (θ) approaches *one* and the value of cos (θ) approaches *zero*, the value of cot (θ) also approaches zero. As the value of sin (θ) approaches zero and the value of cos (θ) approaches one, however, the value of cot (θ) tends to infinity. The value of cot (θ) when cos (θ) equals one is thus said to be *undefined*. As mentioned, the *tangent* function is the quotient of the sine and cosine functions. Since the cotangent function is the *reciprocal* of the tangent function, we can also express the cotangent function in terms of the sine and cosine function:

$$Cot(0) - \frac{cos(\theta)}{sin(\theta)}$$

Obviously, since the cotangent function *is* the reciprocal of the tangent function, it can be expressed in terms of the tangent function as:

$$Cot(\theta) = \frac{1}{tan(\theta)}$$

And finally, since the cosine and sine functions are represented by the x and y coordinates respectively of point P (i.e. the point at which the terminal side of angle θ intersects the circumference of the unit circle), we can also conclude that:

$$Cot(\theta) = \frac{x}{y}$$

We can also define the cotangent function in terms of a right-angled triangle.

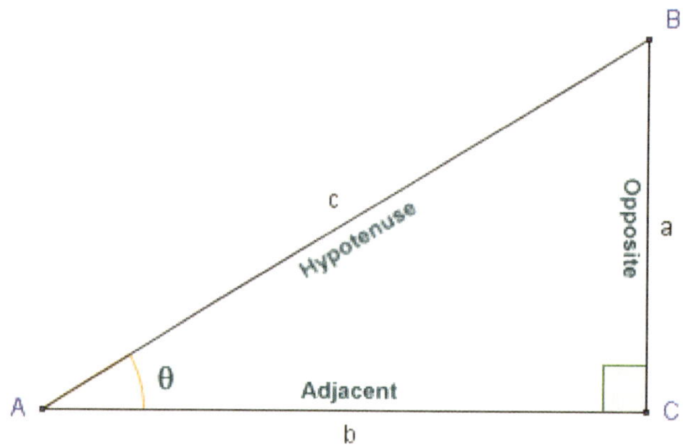

In triangle ABC, $Cot(\theta) = b/a$

In right-angled triangle ABC, the tangent of angle θ is the quotient of the *opposite* and the *adjacent* (i.e. the sides labeled a and b respectively). Since the cotangent function is the *reciprocal* of the tangent function, and then the cotangent of angle θ will be the quotient of the *adjacent* and the *opposite*. This gives us:

$$cot(\theta) = \frac{\cos(\theta)}{\sin(\theta)} = \frac{adjacent}{opposite} = \frac{b}{a}$$

For those interested in such things, we have used *Microsoft Excel* to produce our own table of cotangent values for angles ranging from *zero degrees* (0°) up to *three hundred and sixty degrees* (360°) in increments of one tenth of a degree.

A significant range of angles centered around *ninety degrees* (90°) and *two hundred and seventy degrees* (270°) for which the value returned by the cotangent function is quite small and changes relatively slowly. For angles close to *zero degrees*(0°), *one hundred and eighty degrees* (180°) or *three hundred and sixty degrees* (360°) however, the value returned by the secant function rapidly gets bigger. This is to be expected, since the value of tan (θ) is very small for these angles, so its *reciprocal* will be correspondingly large. In fact, the value returned by the cotangent function for an angle of zero degrees, one hundred and eighty degrees or three hundred and sixty degrees is considered to be *undefined*, since the equation $cot\ (\theta) = {}^{1}/_{tan\ (\theta)}$ will involve division by zero. In fact, it will apply to *any* angle for which the tangent value is zero. We present below the graph of the cotangent function for angles in the range *zero* to *seven hundred and twenty degrees* (720°).

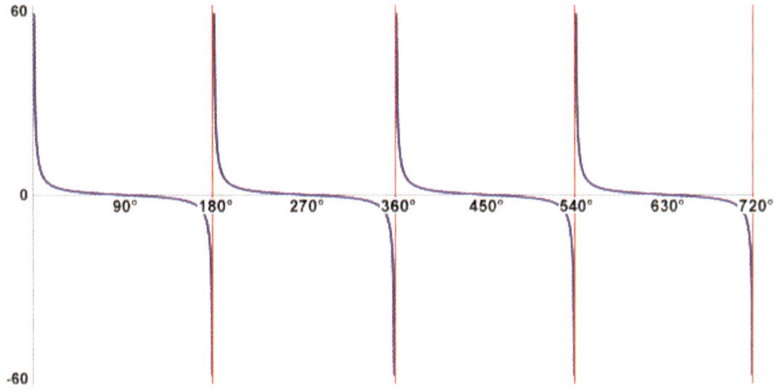

A graph of the cotangent function for angles in the range 0° to 720°

Like the tangent function, the values of the cotangent function will vary between zero and infinity (or more correctly, between zero and a very large *undefined* value). Because the cotangent function is the reciprocal of the tangent function, the cotangent value will be undefined when the tangent value is zero, and zero when the tangent value is undefined. As you can see by studying the table of cotangent values and the graph of the cotangent function, the cotangent value is negative for all angles for which the sine and cosine have a *different sign* (i.e. in *Quadrant II* and *Quadrant IV*), and positive for all angles for which the sine and cosine have the *same sign* (i.e. in *Quadrant I* and *Quadrant III*).

Secant Function

In a right triangle, the secant of an angle is the length of the hypotenuse divided by the length of the adjacent side. In a formula, it is abbreviated to just 'sec'.

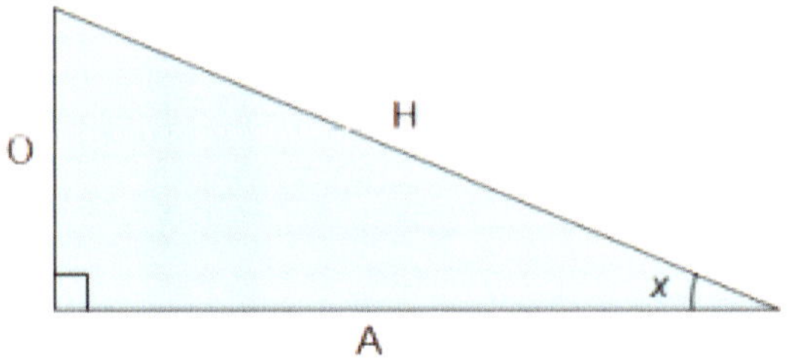

$$\sec x = \frac{H}{A}$$

Of the six possible trigonometric functions, secant, cotangent, and cosecant, are rarely used. In fact, most calculators have no button for them, and software function libraries do not include them.

They can be easily replaced with derivations of the more common three: sin, cos and tan. Secant can be derived as the reciprocal of cosine:

$$\sec x = \frac{1}{\cos x}$$

The Inverse Secant Function - Arcsec

For every trigonometry function such as sec, there is an inverse function that works in reverse. These inverse functions have the same name but with 'arc' in front. So the inverse of sec is arcsec etc. When we see "arcsec A", we interpret it as "the angle whose secant is A".

sec 60 = 2.000	Means: The secant of 60 degrees is 2.000
arcsec 2.0 = 60	Means: The angle whose secant is 2.0 is 60 degrees.

Sometimes written as asec or sec^{-1}

Large and Negative Angles

In a right triangle, the two variable angles are always less than 90°. But we can in fact find the secant of any angle, no matter how large, and also the secant of negative angles.

Graph of the Secant Function

Because the secant function is the reciprocal of the cosine function, it goes to infinity whenever the cosine function is zero.

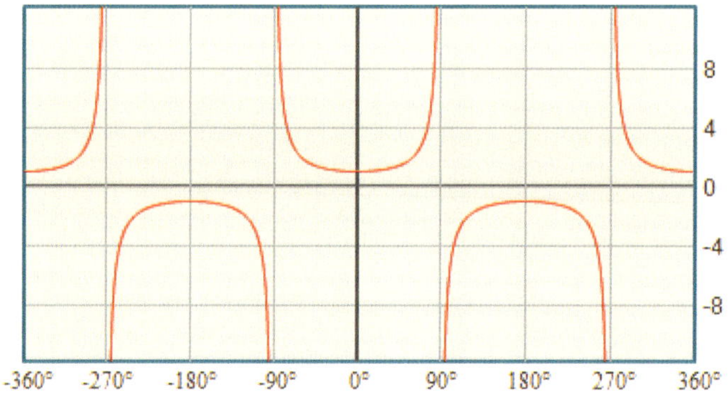

The Derivative of Sec(x)

In calculus, the derivative of *sec(x)* is *sec(x) tan(x)*. This means that at any value of *x*, the rate of change or slope of *sec(x)* is *sec(x) tan(x)*.

Cosecant Function

The *cosecant function* (usually abbreviated as csc) is the reciprocal function of the *sine function*. As with the other trigonometric functions in modern use, the cosecant function was certainly known to Arab scholars, and tables of cosecant values are known to have been in existence by the tenth century CE. In order to clarify exactly how the cosecant function and the sine function are related, consider the diagram.

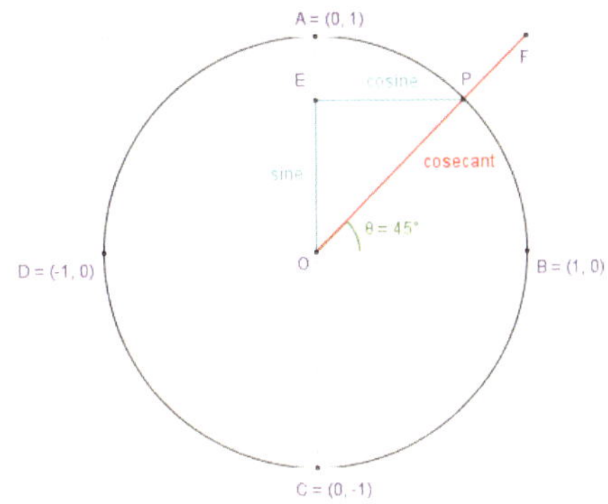

The cosecant as a line segment of the unit circle

Like the other trigonometric functions, the cosecant can be represented as a line segment associated with the *unit circle*. The diagram shows the cosecant for an angle of rotation θ of *forty-five degrees* (measured anti-clockwise from the positive x-axis). Line segment AF lies on the line that is tangent to the circle at point A. Line segment OF (shown in red) represents the cosecant, and is an extension of line segment OP. Remember that point P is a point on the circumference of the circle whose x and y coordinates represent the value of cos (θ) and sin (θ) respectively (the line segments representing the sine and cosine are also shown). that as the value of sin (θ) approaches *one*, so does the value of csc (θ). As the value of sin (θ) approaches zero, however, the value of csc (θ) tends to infinity. The value of csc (θ) when sin (θ) equals zero is thus said to be *undefined.*

Obviously, since the cosecant function is the reciprocal of the sine function, it can be expressed in terms of the sine function as:

$$csc(\theta) = \frac{1}{\sin(\theta)}$$

Since the sine is represented by the y coordinate of point P (i.e. the point at which the terminal side of angle θ intersects the circumference of the unit circle), we can also conclude that:

$$csc(\theta) = \frac{1}{y}$$

We can also define the cosecant in terms of a right-angled triangle.

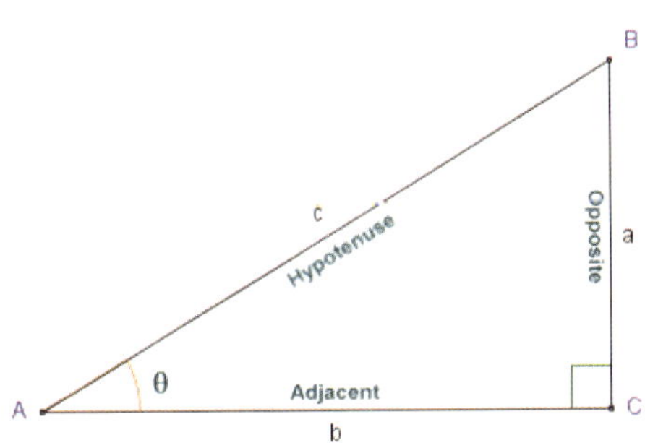

In triangle ABC, csc(θ) = c/a

In right-angled triangle ABC, the sine of angle θ is the quotient of the *opposite* and the *hypotenuse* (i.e. the sides labelled a and c respectively). Since the cosecant function is the *reciprocal* of the sine function, and then the cosecant of angle θ will be the quotient of the *hypotenuse* and the *opposite*. This gives us:

$$csc(\theta) = \frac{1}{\sin(\theta)} = \frac{hypotenuse}{opposite} = \frac{c}{a}$$

A significant range of angles centered around ninety degrees for which the value returned by the cosecant function is quite small and changes relatively slowly. For angles close to either *zero degrees* (0°) or *one hundred and eighty degrees* (180°) however, the value returned by the cosecant function is much bigger. This is to be expected, since the value of sin (θ) will be very small for these angles, so its *reciprocal* will be correspondingly large. In fact, the value returned by the cosecant function for an angle of either zero degrees or one hundred and eighty degrees is considered to be *undefined*, since the equation $csc(\theta) = {}^{1}/_{sin\ (\theta)}$ will involve division by zero. The same is true for an angle of *three hundred and sixty degrees* (360°). In fact, it will apply to any angle for which the sine value is zero. We present below the graph of the cosecant function for angles in the range *zero degrees* to *seven hundred and twenty degrees* (720°).

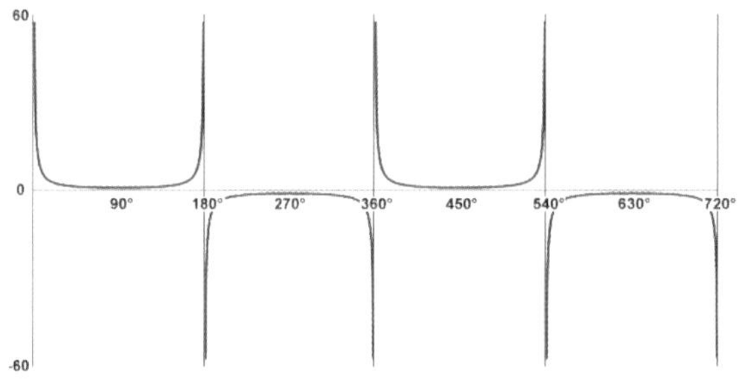

A graph of the cosecant function for angles in the range 0° to 720°

Because the absolute value of the sine of an angle will never exceed *one*, the reciprocal of the sine of an angle will never be less than one. So, just as the cosecant value is undefined for any angle for which the sine is *zero*, it will always be *one* for any angle for which the sine is one, and *minus one* (-1) for any angle for which the sine is minus one. In fact, as you can see from both the table of cosecant values and the graph of the cosecant function, the cosecant value is negative for *all* angles for which the sine is negative. Like the graph of the tangent function, the graph of the cosecant function can include some very large positive and negative peaks, as the value of sin (θ) tends to zero. Unlike the tangent function however, the period of the cosecant function is *three hundred and sixty degrees* (360°), or 2π. Like the sine function, the cosecant function returns positive values for angles in *Quadrants I and II*, and negative values for angles in *Quadrants III and IV*.

References

- Cotangent-function, trigonometry, mathematics: technologyuk.net, Retrieved 14 April 2018
- Joseph, George G., The Crest of the Peacock: Non-European Roots of Mathematics, 2nd ed. Penguin Books, London. (2000). ISBN 0-691-00659-8Trigonometric-function, science: britannica.com, Retrieved 16 May 2018
- Maor, Eli, Trigonometric Delights, Princeton Univ. Press. (1998). Reprint edition (February 25, 2002): ISBN 0-691-09541-8Sine-function, trigonometry, mathematics: technologyuk.net, Retrieved 28 March 2018
- Basic-trigonometric-functions: brilliant.org, Retrieved 19 June 2018

3

Graphing and Periodic Trigonometric Functions

Trigonometric functions are periodic, which means that they repeat their values in regular periods. The understanding of trigonometric functions is incomplete without a study of the plotting of their graphs on a XY plane. This chapter provides a detailed discussion of the graphing of periodic trigonometric functions.

Graphing Trigonometric Functions

Graphs are useful because they can take complicated information and display it in a simple, easy-to-read manner. Graphing trigonometric functions on the coordinate system is defined as the x values will be the angles (in radians – that's the way it's done), and the y values will be the trig value (like sin, cos, and tan).

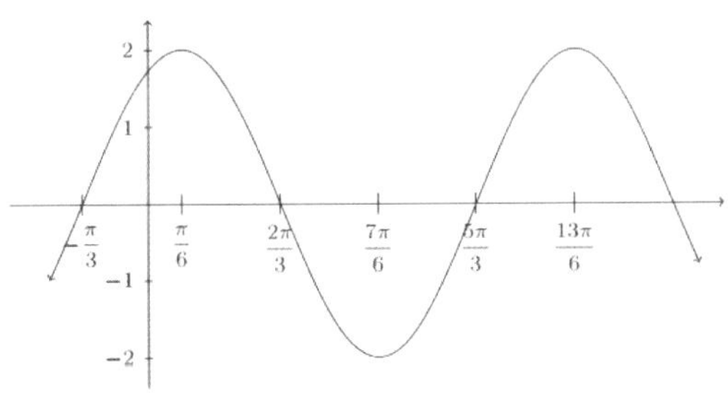

The sin and cos (and csc and sec) functions start repeating after 2π radians, and it turns out that the tan and cot functions start repeating again after only π radians. The reason tan (and cot) repeat after only π radians is because, when dividing sin and cos to get tan, we get the same values in quadrants III as IV as we do for quadrants I and II. (Try this with the Unit Circle).

A complete repetition of the pattern of the function is called a cycle and the period is the horizontal length of one complete cycle. Thus the period of the sin, cos, csc, and sec graphs is 2π radians, and the period for the tan and cot graphs is π radians.

Table of Trigonometric Parent Functions

Before we go into more detail of each of the trig functions, here is a table of the Trigonometric Parent Functions. Note that when the domain can't be certain values, we have asymptotes at those values of x.

Starting and stopping points may be changed, as long the graph covers one complete cycle (period).

Trig Function T-charts

$y = sin(x)$

x	y
0	0
$\frac{\pi}{2}$	1
π	0
$\frac{3\pi}{2}$	−1
2π	0

$y = cos(x)$

x	y
0	1
$\frac{\pi}{2}$	0
π	−1
$\frac{3\pi}{2}$	0
2π	1

$y = tan(x)$

x	Y
$-\frac{\pi}{2}$	undef
—	−1
0	0
$\frac{\pi}{4}$	1
$\frac{\pi}{2}$	undef

$y = csc(x)$

x	Y
0	undef
$\frac{\pi}{2}$	1
π	undef
$\frac{3\pi}{2}$	−1
2π	undef

$y = sec(x)$

x	Y
$-\frac{\pi}{2}$	undef
0	1
$\frac{\pi}{2}$	undef
π	−1
$\frac{3\pi}{2}$	undef

$y = cot(x)$

x	y
0	undef
$\frac{\pi}{4}$	1
$\frac{\pi}{2}$	0
$\frac{3\pi}{4}$	−1
π	undef

Parent Function	Graph	Parent Function	Graph
$y = \sin(x)$ Odd Domain: $(-\infty,\infty)$ Range: $[-1,1]$ Period : 2π Zeros: $(\pi k,0)$ k $\in$ Integers	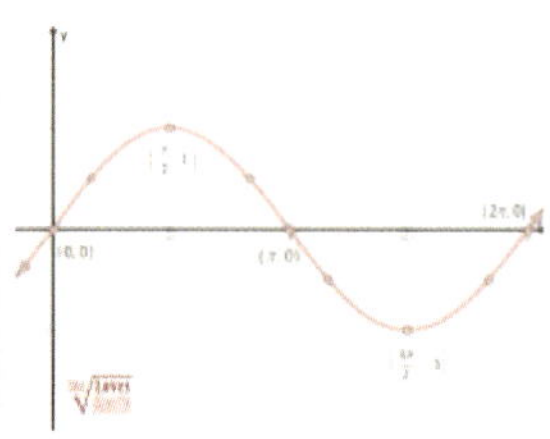	$y = \csc(x)$ Odd Domain: $x \neq \pi k$ Range: $(-\infty,-1]\cup[1,\infty)$ Asymptotes: $x = \pi k$ Period: 2π Zeros: None	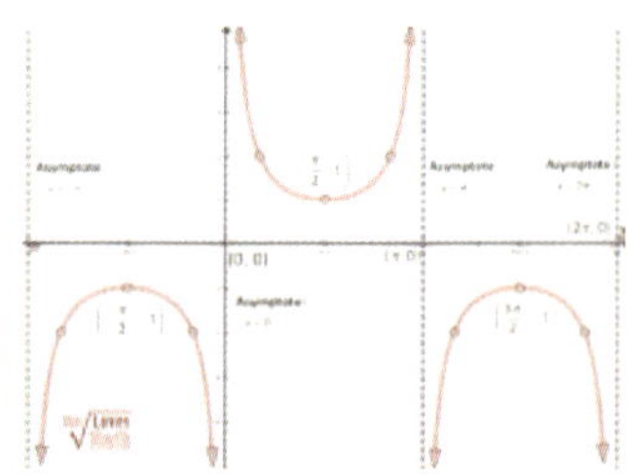
$y = \cos(x)$ Even Domain: $[-\infty,\infty]$ Range: $[-1,1]$ Period: 2π Zeros: $\left(\frac{\pi}{2}+\pi k,0\right)$	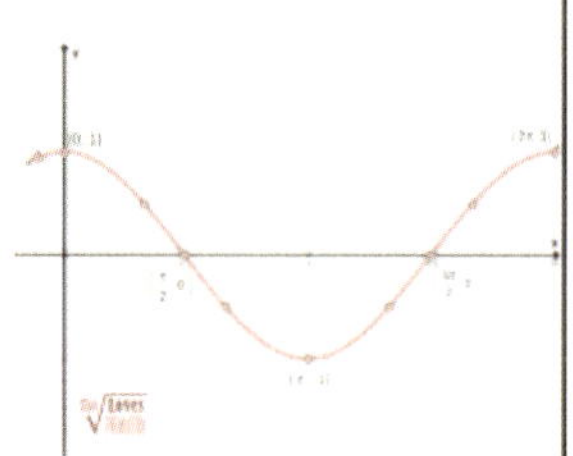	$y = \sec(x)$ Even Domain: $x \neq \frac{\pi}{2}+\pi k$ Range: $(-\infty,-1]\cup[1,\infty)$ Asymptotes: $x = \frac{\pi}{2}+\pi k$ Period: 2π Zeros: None	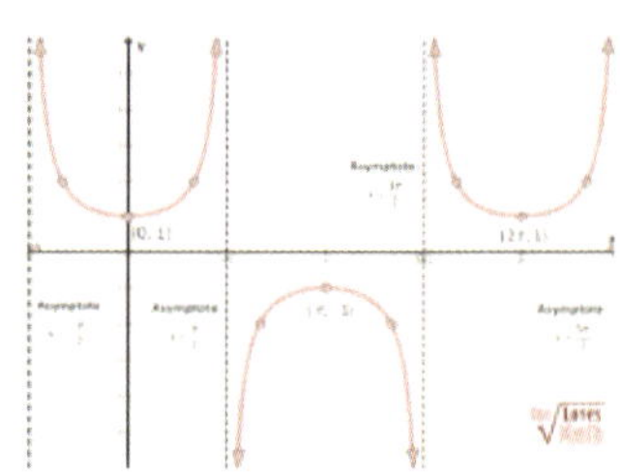

$y = \tan(x)$ Odd Domain: $x \neq \frac{\pi}{2} + \pi k$ Range: $(-\infty, \infty)$ Asymptotes: $x = \frac{\pi}{2} + \pi k$ Period: π Zeros: $(\pi k, 0)$	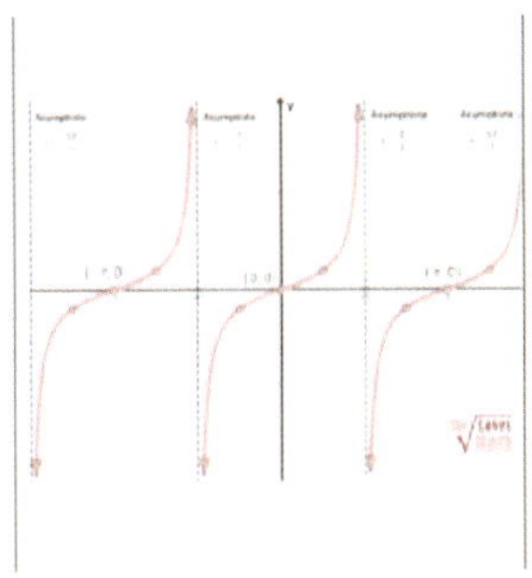	$y = \cot(x)$ Odd Domain: $x \neq \pi k$ Range: $(-\infty, \infty)$ Asymptotes: $x = \pi k$ Period: π Zeros: $\left(\frac{\pi}{2} + \pi k, 0\right)$	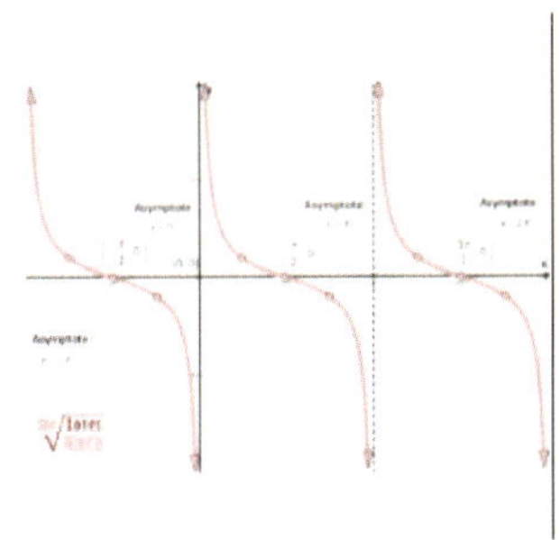

Periodic Trigonometric Functions

Period means the time interval between two waves. Periodic function is a function that repeats its values at regular intervals or periods. In other words, a function which repeats its values after every particular interval, is known as a periodic function. This particular interval is termed as the period of that function.

A function f is said to be periodic with period m, if we have f (x + m) = f (x), For every m > 0.

It means that the function f(x) possess same values after an interval of “m”. One can say that the function f repeats all its values after every interval of “m”.

For example - The sine function i.e. sin x has a period 2 ππ because 2 ππ is the smallest number for which sin (x + 2π) − sin x, for all x.

We may also calculate the period using the formula derived from the basic sine and cosine equations. The period for function y = A sin(B x - c) and y = A cos(B x - c) is $\frac{2\pi}{B}$ radians. The reciprocal of period of a function is equal to its frequency. Frequency is defined as the number of cycles completed in one second. If the period of a function is denoted by P and f be its frequency, then -

$$f = \frac{1}{p}.$$

Fundamental Period of a Function

The fundamental period of a function is the period of the function which are of the form $f(x+k) = f(x)$, then k is called the period of the function and the function f is called a periodic function.

Let us define the function h(t) on the interval [0,2] as follows:

$$h(t)=\begin{cases} 0 & \text{if } x=0\le t\le 1 \, or \text{ if } \frac{5}{3}\le t\le 2 \\ 3t-1 & \text{if } x=\frac{1}{3}\le t\le \frac{2}{3} \\ 1 & \text{if } x=\frac{2}{3}\le t\le \frac{4}{3} \\ -3t+5 & \text{if } x=\frac{4}{3}\le t\le \frac{5}{3} \end{cases}$$

If we extend the function h to all of R by the equation, $h(t+2)=h(t)$ => h is periodic with the period 2. The graph of the function is shown below.

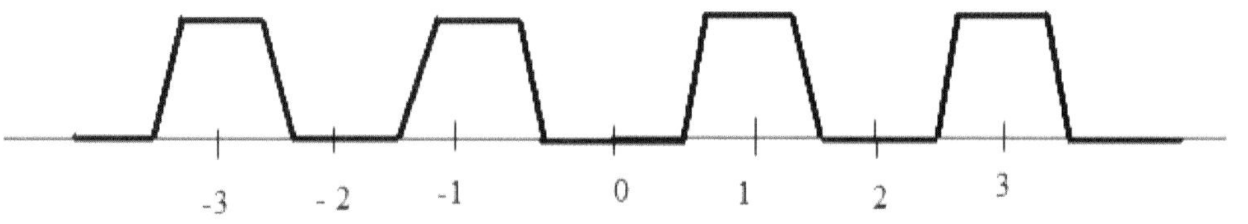

Finding the Period of a Function

If a function repeats over at a constant period we say that is a periodic function. Basically it is represented like f(x) = f(x + p), p is the real number and this is the period of the function. Period means the time interval between the two occurrences of the wave. To find the period of the periodic function we have to use the following formula, Where Period = 2pb, here b is the co - efficient of x

Sine and cosine functions have the forms of a periodic wave:

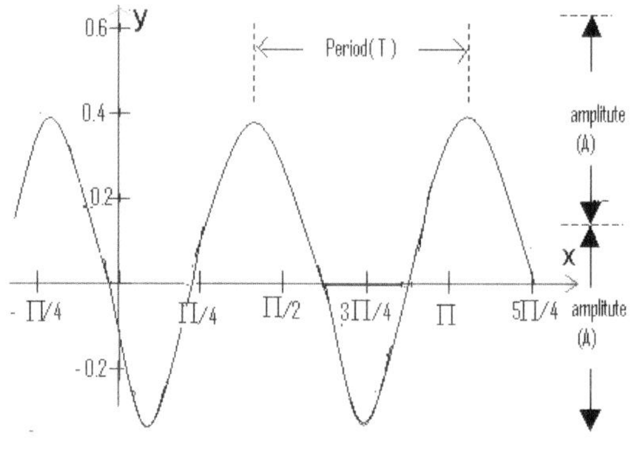

Period: It is represented as "T", Period is the distance among two repeating points on the graph function.

Amplitude: It is represented as "A" It is the distance between the middle point to highest or lowest point on the graph function.

$$\sin(a\theta)=\frac{2\pi}{a}$$

and $\cos(a\theta) = \frac{2\pi}{a}$

Example: Find the period of the given periodic function. Where f(x) = 9sin (6px7 + 5)

Solution:

Given periodic function is f(x) = 9sin (6px7+ 5)

To find the period we have the formulas

Period = 2pb

Where period of the periodic function = 2p (6p7) = 146 = 73

Example: Find the period of the periodic function f(x) . Where f(x) = 9 Cos x

Solution:

The given periodic function is f(x) = 9 Cos x

To find the period, we have the formula.

Period = $\frac{2\pi}{b}$

Where period of the periodic function = $2p \times 1 = 2p$

As we are aware that sin (2π + x) = sin x and cos (2π + x) = cos x.., we see that the periods of sine and cosine functions are 2π.

Also, tan (π + x) = tan x, hence the period of tan x is π

Let us graph the primary trigonometric functions.

The following graph shows the function y = sin x

Let us find the co-ordinates of the points to graph.

x	-2 π	-3 π/2	- π	- π/2	0	π/2	π	3 π/2	2π	5 π/2
y	1	1	0	-1	0	1	0	-1	0	1

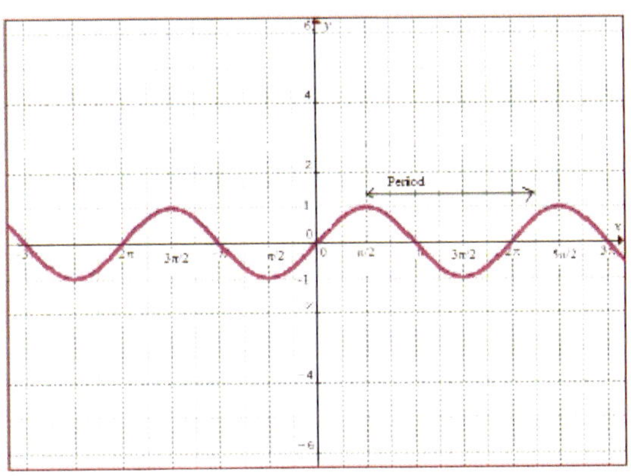

Period = 2 π

Axis: y = 0 [x-axis]

Amplitude : 1

Maximum value = 1

Minimum value = -1

domain = { x : x ∈ R }

Range = [-1, 1]

The following graph shows the trigonometric function **y = cos x**

Let us prepare the table to values

x	-2 π	-3 π/2	- π	- π/2	0	π/2	π	3 π/2	2π	5 π/2
y	1	0	-1	0	1	0	-1	0	1	0

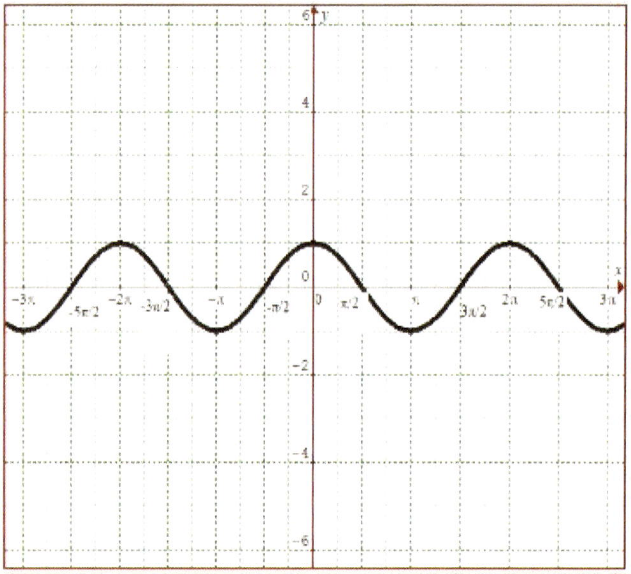

Period: 2 π

Axis: y = 0

Amplitude = 1

Maximum value = 1

Minimum value = -1

Domain: { x : x ∈ R }

Range = [-1, 1]

Period of a Sine Function

If we have a function f(x) = sin (xs), where s > 0, then the graph of the function makes s complete cycles between 0 and 2π and each of the function have the period, $p = \frac{2\pi}{s}$

Now, lets discuss about some examples based on sin function:

Let us discuss the graph of y = sin 2x

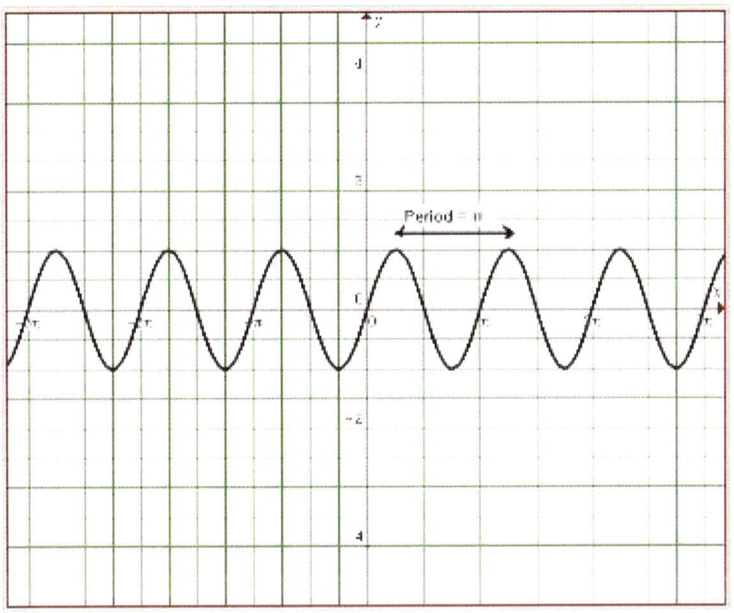

Period = π

Axis: y = 0 [x- axis]

Amplitude: 1

Maximum value = 1

Minimum value = -1

Domain: { x : x ∈ R }

Range = [-1, 1]

Period of a Trig Function

The distance between the repetition of any function is called the period of function. For a trigonometric function, the starting point.

In general we have three basic trigonometric functions like sin, cos and tan function, having 2π, 2π and π period respectively.

Example: Graph of y = 4 sin(x/2)

Solution:

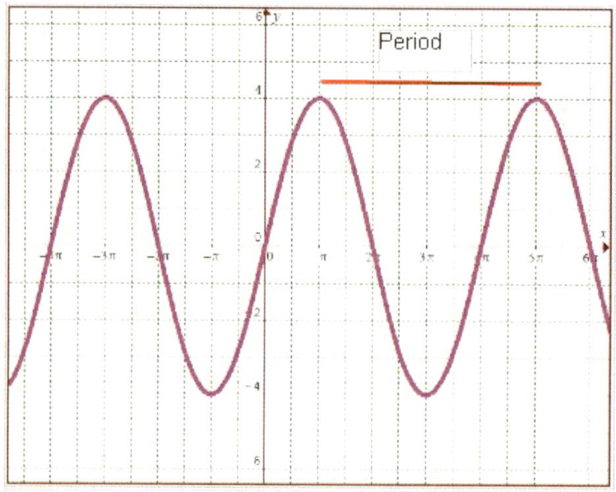

Period = 4π

Axis: $y = 0$ [x-axis]

Amplitude: 4

Maximum value = 4

Minimum value = -4

Domain: $\{ x : x \in R \}$

Range = [-4, 4]

Example: Graph of $y = 4 \sin 2x$

Solution:

x	$-\pi$	$-3\pi/4$	$-\pi/2$	$-\pi/4$	0	$\pi/4$	$\pi/2$	$3\pi/4$	π
y	0	4	0	-4	0	4	0	-4	0

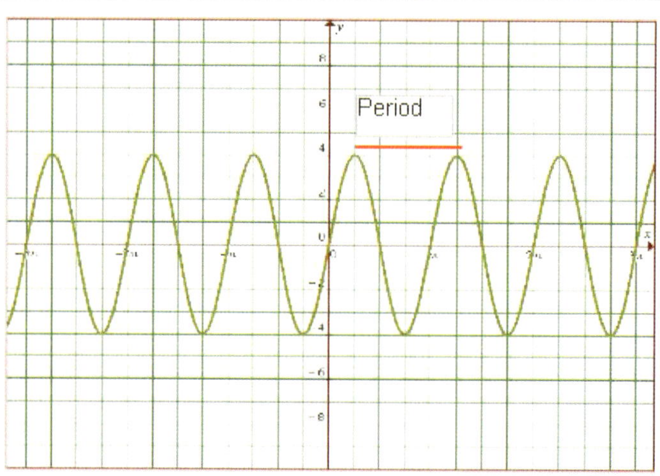

Period = π

Axis: y = 0 [x-axis]

Amplitude: 4

Maximum value = 4

Minimum value = -4

Domain: { x : x ∈ R }

Range = [-4, 4]

From the above examples, we observe that the period and amplitude of the function changes as there is change in the numerical co-efficient and the angle [argument].

Let us tabulate the argument and the amplitude and compare.

Function	Amplitude	Period
y = sin x	1	2π
y = 4sin 2x	4	π
y = 4 sin(x/2)	4	4π

Hence, as the argument increases the period decreases and as the argument decreases the period increases.

Let us discuss the following general formula.

y = f(x) = a sin(k (x - b) + c and

y = f(x) = a cos (k(x - b)) + c

Since the primary functions sin x and cos x are of period 2π , from the above general equation, the period = $\frac{2\pi}{|k|}$

Here, 'a' represents the amplitude

x = b, represents the horizontal shift

and y = c represents the axis.

Period of a Tangent Function

If we have a function f(x) = tan (xs), where s > 0, then the graph of the function makes s complete cycles between $-\frac{\pi}{2}$,0 and $\frac{\pi}{2}$ and each of the function have the period of p = $\frac{\pi}{s}$

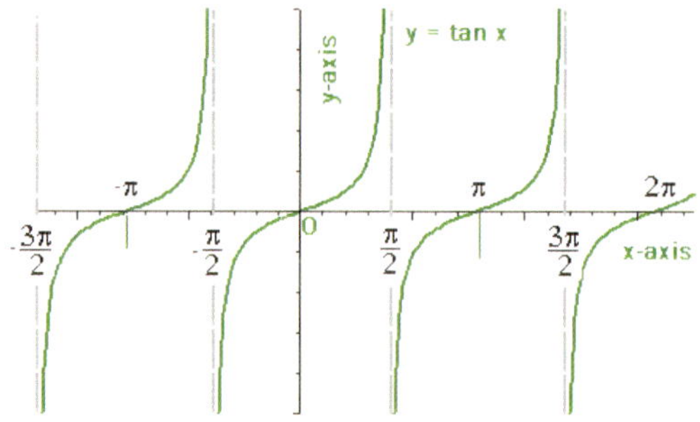

Example: Determine the period of the function $f(\theta)=\tan\left(\frac{3\theta}{2}\right)$

Solution:

For the trigonometric tan function the period p is given as follows:

$$\text{p}=\frac{\pi}{s}=\frac{\pi}{\frac{3}{2}}=\frac{2\pi}{3}$$

Below is the graph of the trigonometric function y = tan x, where x = 3θ/2

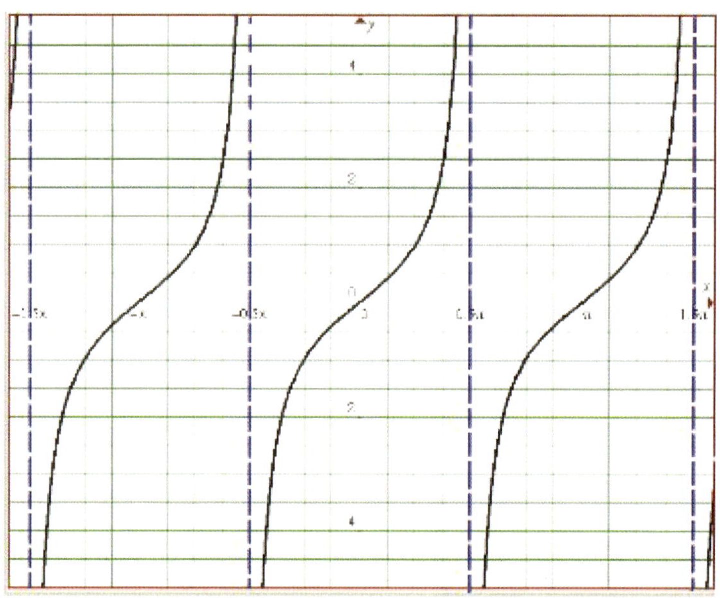

Period = —

Axis y = 0

Amplitude: undefined

No maximum or minimum values.

Period of a function examples

Given below are some of the examples on periodic functions.

Example: What is the period of the following periodic function?

Where f(x) = 6 sin 5x

Solution:

The given periodic function is f(x) = 6 sin 5x

We have the formula for the period of a function.

$$\text{period} = \frac{2\pi}{b}$$

Where period of the periodic function = $\frac{2\pi}{5}$

Example: The frequency of the periodic function is 2phi. Find the period of the periodic function.

Solution:

The frequency of the periodic function will be $f = \frac{1}{p}$

So, period $p = \frac{1}{f}$

$$= \frac{1}{2\pi}$$

Example: Find the period, amplitude, horizontal shift, vertical shift and the equation of the axis for the following functions.

y = 3 cos(5x - 5π) + 2

Solution:

We have y = 3 cos(5x - 5 π) + 2 = 3cos(5(x - π))+ 2

Comparing this with the general equation,

y = a cos (k(x - b)) + c

we get a = 3, k = 5, b = π, c = 2

Hence, the Period of the function is $\frac{2\pi}{|k|} = \frac{2\pi}{5}$

Amplitude of the function is a = 3

As b = π, the graph is shifted π units to the right.

As c = 2, the equation of the axis is y = 2.

We can draw the graph as follows.

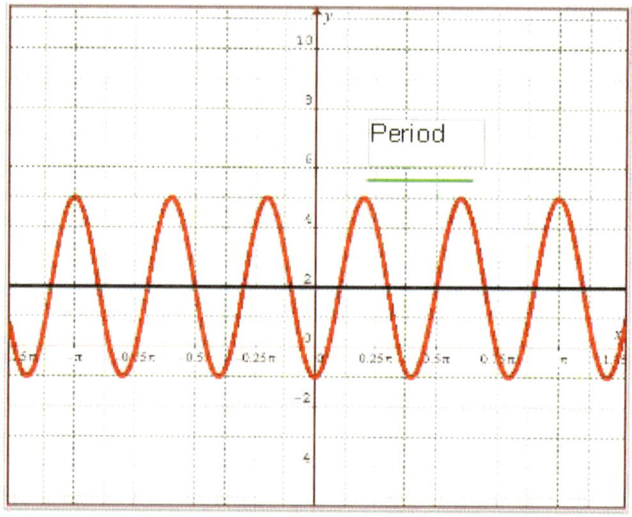

Example: The following trigonometric function have the basic function y = sin x. Determine the equation of each of the function a. The graph of the trigonometric function has period π and amplitude 12. The equation of the axis is y = -6.

Solution:

From the given data, a = 12, b = π, c = -6

Period = $2\pi/|k|$

k = $2\ \pi/\text{period} = 2\ \pi/\pi = 2$

The general equation is, y = a sin(k (x-b) + c

Hence, our required equation is, y = 12 sin(2x) - 6

The graph of the function will be as below.

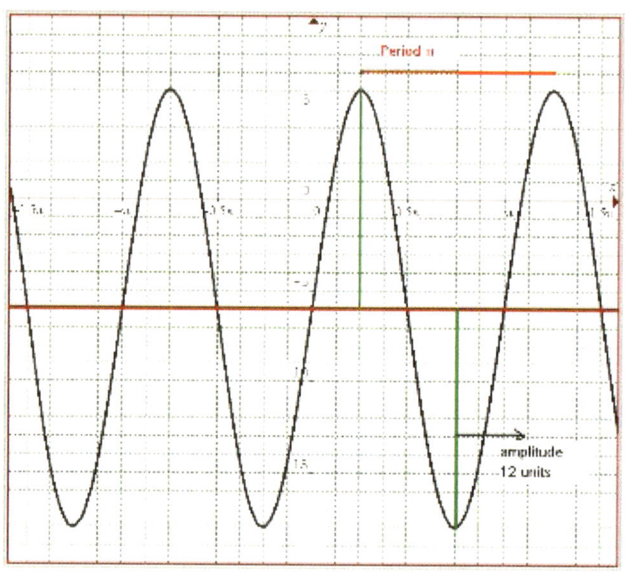

Graph of y = sin x

y = sin x is periodic function. The period of y = sin x is 2π. Therefore, we will draw the graph of y = sin x in the interval [-π, 2π].

For this, we need to take the different values of x at intervals of 10°. Then by using the table of natural sines we will get the corresponding values of sin x. Take the values of sin x correct to two place of decimal. The values of sin x for the different values of x in the interval [-π, 2π] are given in the following table.

We draw two mutually perpendicular straight lines XOX' and YOY'. XOX' is called the x-axis which is a horizontal line. YOY' is called the y-axis which is a vertical line. Point O is called the origin.

Now represent angle (x) along x-axis and y (or sin x) along y-axis.

Along the x-axis: Take 1 small square = 10°.

Along the y-axis: Take 10 small squares = 1 unity.

Now plot the above tabulated values of x and y on the co-ordinate graph paper. Then join the points by free hand. The continuous curve obtained by free hand joining is the required graph of y = sin x.

Steps to draw the graph of y = c sin ax.

Steps I: Obtain the values of a and c.

Step II: Draw the graph of y = sin x and mark the points where y = sin x crosses x-axis.

Step III: Divide the x-coordinate of the points where y = sin x crosses x-axis by a and mark maximum and minimum values of y = c sin ax as c and –c on y-axis.

The graph obtained is the required graph of y = c sin ax.

Properties of y = sin x

(i) The graph of the function y = sin x is continuous and extends on either side in symmetrical wave form.

(ii) Since the graph intersects the x-axis at the origin and at points where x is an even multiple of 90°, hence sin x is zero at x = nπ where n = 0, ±1, ±2, ±3, ±4, ...

(iii) The ordinate of any point on the graph always lies between 1 and - 1 i.e., $-1 \le y \le 1$ or ,$-1 \le \sin x \le 1$ hence, the maximum value of sin x is 1 and its minimum value is - 1 and these values occur alternately at $\frac{\pi}{2}, \frac{3\pi}{2}, \frac{5\pi}{2}, ...$ i. e., at $x = (2n + 1)\frac{\pi}{2}$, where n = 0, ±1, ±2, ±3, ±4, ...

(iv) Since the function y= sin x is periodic of period 2π, hence the portion of the graph between 0 to 2π is repeated over and over again on either side.

Solved example to sketch the graph of y = sin x:

Sketch the graph of y = 2 sin 3x.

Solution:

To obtain the graph of y = 2 sin 3x we first draw the graph y = sin x in the interval [0, 2n] and then divide the x-coordinates of the points where it crosses x-axis by 3. The maximum and minimum values are 2 and -2 respectively.

Graph of Sine *x* - with Varying Amplitudes

We start with $y = \sin x$.

It has amplitude =1 and period=2π.

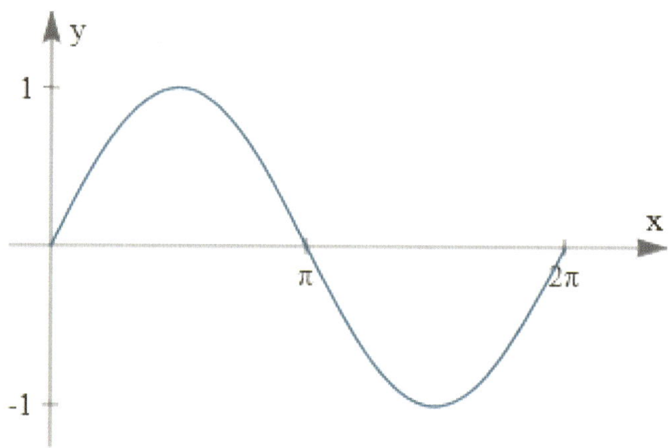

The graph of $y = \sin(x)$ for $0 \le x \le 2\pi$

Now let's look at the graph of $y = 5\sin x$.

This time we have amplitude $= 5$ and period $= 2\pi$. (I have used a different scale on the y-axis.)

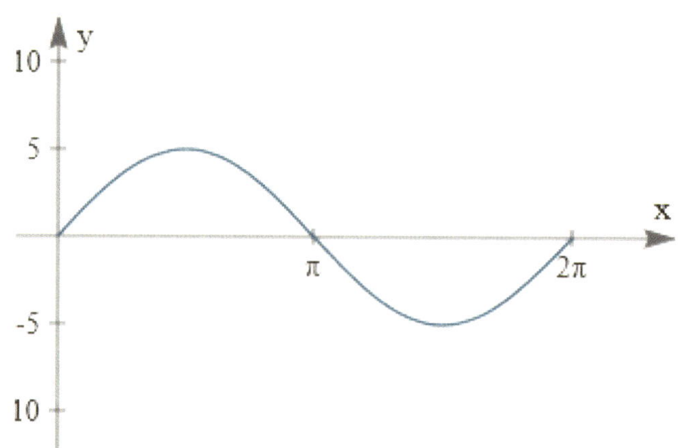

The graph of $y = 5\sin(x)$ for $0 \le x \le 2\pi$

And now for y = 10 sin x.

Amplitude = 10 and period = 2π.

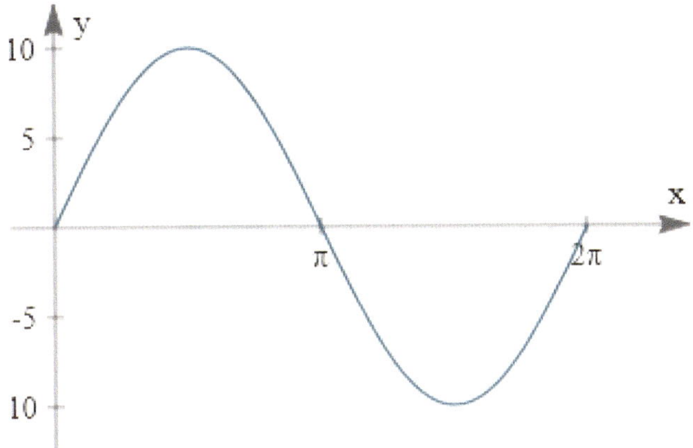

The graph of $y = 10\sin(x)$ for $0 \le x \le 2\pi$

For comparison, and using the same y-axis scale, here are the graphs of

$p(x) = \sin x,$

$q(x) = 5\sin x$ and

$r(x) = 10\sin x$

on the one set of axes.

Note that the graphs have the same period (which is 2π) but different amplitude.

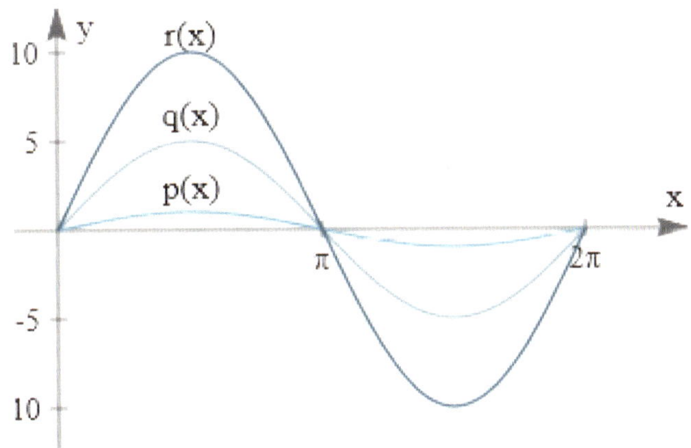

The graphs of $p(x), q(x)$, and $r(x)$ for $0 \le x \le 2\pi$

Graph of y = cos x

To draw the graph of the cosine function divide the unit circle and x-axis of a Cartesian coordinate system the same way as when drawing the sine function.

The abscissas of the ending points of arcs x, of the unit circle, are now the ordinates of the corresponding points $P\ (x,\ cos\ x)$ of the graph, as shown in the figure.

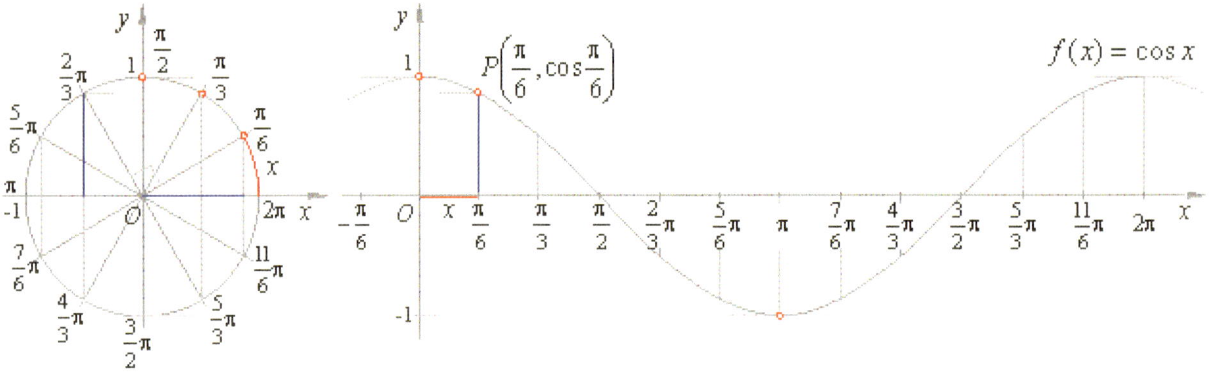

Since $\cos x = \sin(x + \pi / 2)$ that is, the cosine of an arc x equals the sine of the same arc increased by $\pi / 2$.

Therefore, the graphs of the cosine function correspond to the graph of the sine function translated in the negative direction of the x-axis by $\pi / 2$.

Thus for example, $\cos \pi / 6 = \sin(\pi / 6 + \pi / 2) = \sin 2\pi / 3$ as shows the above graph.

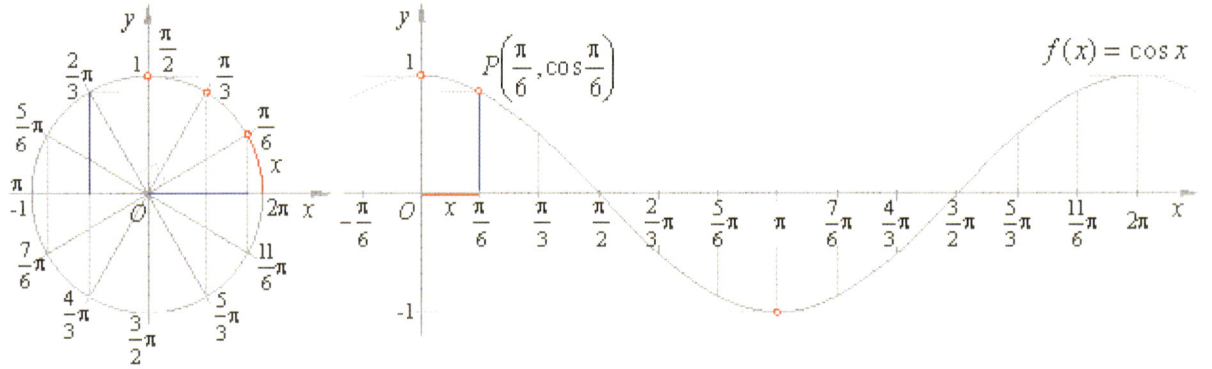

Properties of the cosine function

- Domain, $x \in R$.
- Range, $-1 \leq y \leq 1$.
- Zeros of the function, $x = \pi / 2 + k\pi, k \in Z$
- Abscissas of maximums, $x = k.2\pi,\ k \in Z$ and minimums $x = (2k+1) \cdot \pi, k \in Z$.
- Parity and periodicity, the cosine is even function since $f(-x) = \cos(-x) = \cos x = f(x)$.

The identity $\cos(x + k \cdot 2\pi) = \cos x, k \in Z$

shows that the cosine is periodic function with the period $P = 2\pi$

Behavior of the Cosine Function

The table shows behavior of the cosine function while the arc x of the unit circle increases passing through all the values from the period.

maximum minimum maximum

x	0	↗	$\frac{\pi}{2}$	↗	π	↗	$\frac{3\pi}{2}$	↗	2π
$\cos x$	1	↘	0	↘	-1	↗	0	↗	1

zeros

Graph of y = tan x

y = tan x is periodic function. The period of y = tan x is π. Therefore, we will draw the graph of y = tan x in the interval [-π, 2π].

For this, we need to take the different values of x at intervals of 10°. Then by using the table of natural tangent we will get the corresponding values of tan x. Take the values of tan x correct to two place of decimal. The values of tan x for the different values of x in the interval [-π, 2π] are given in the following table.

We draw two mutually perpendicular straight lines XOX' and YOY'. XOX' is called the x-axis which is a horizontal line. YOY' is called the y-axis which is a vertical line. Point O is called the origin.

Now represent angle (x) along x-axis and y (or tan x) along y-axis.

Along the x-axis: Take 1 small square = 10°.

Along the y-axis: Take 10 small squares = 1 unity.

Now plot the above tabulated values of x and y on the co-ordinate graph paper. Then join the points by free hand. The continuous curve obtained by free hand joining is the required graph of y = tan x.

Properties of y = tan x

(i) The tangent-graph is not a continuous curve, but consists of infinite separate branches parallel to one another, the points of discontinuities are at $x = (2n + 1)\frac{\pi}{2}$ where n = 0 or any integer.

(ii) As x passes through any point of discontinuity from the left to the right, the value of tan x suddenly changes from (+∞) to (-∞).

(iii) Each branch of the curve approaches continuously the two lines parallel to y- axis at two points of discontinuity of the graph. Such lines are called asymptotes to the curve.

(iv) Since the function y = tan x is periodic of period π hence each branch is simply a repetition of the branch from $-\frac{\pi}{2}$ to $\frac{\pi}{2}$.

The Graph of y = tan x

Sketch $y = \tan x$.

Solution

As we saw above,

$$\tan x = \frac{\sin x}{\cos x}$$

This means the function will have a discontinuity where $\cos x = 0$. That is, when x takes any of the values:

$$x = \ldots, -\frac{5\pi}{2}, -\frac{3\pi}{2}, -\frac{\pi}{2}, \frac{\pi}{2}, \frac{3\pi}{2}, \frac{5\pi}{2}, \ldots$$

It is very important to keep these values in mind when sketching this graph.

When setting up a table of values, make sure you include x-values either side of the discontinuities.

Recall that $-\frac{3\pi}{2} = -4.7124$ and $-\frac{\pi}{2} = -1.5708$. So we take values quite near these discontinuities.

x	-4.7	-4.5	-4	-3.5	-3.14	-1.58	-1.56	-1	0	0.5	1	1.5	1.57
tan x	-80.7	-4.6	-1.2	-0.37	0	108	-92	-1.6	0	0.55	1.6	14.1	1,256

Notice that either side of $-\frac{\pi}{2}$ (our values of -1.58 and -1.56 in the table above), we jump from a large positive number (108), to a small negative number (-92).

If we continue our table, we will get similar values (because this is a periodic graph). So we are ready to sketch our curve.

Graph of $y = \tan x$:

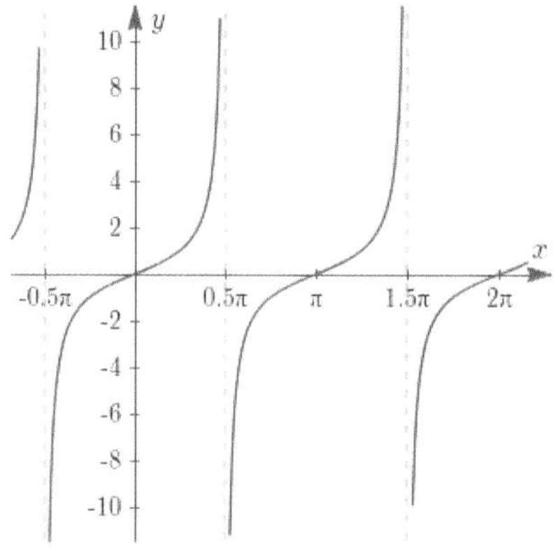

The graph of $y = \tan(x)$ for $-\frac{\pi}{2} \le x \le 2\pi$

Note that there are vertical asymptotes (the gray dotted lines) where the denominator of x has value zero.

Graph of y = cot x

A Cot function graph looks like; notice that the domain is x≠πk, *k* is an integer (where the asymptotes are), and the range is $(-\infty,\infty)$.

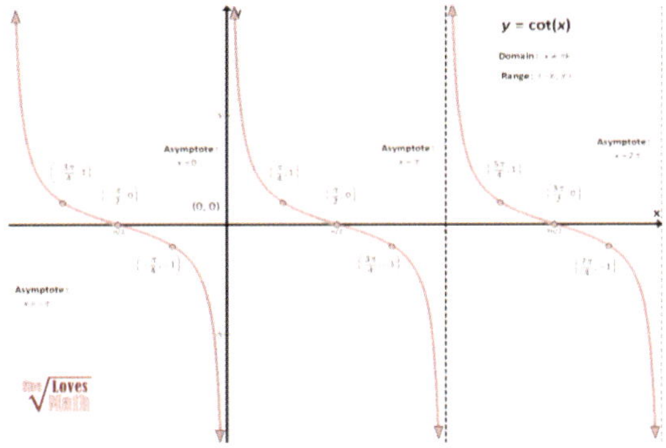

Note the cotangent graph repeats itself (cycles) every π radians (the *x*); this is called the period of the graph. The graph repeats itself one-half of a revolution of the unit circle.

Note also that there is a vertical asymptote every π radians; this is because for multiples of π, we have to divide by 0 to get the tangent; this creates the asymptote. The vertical asymptotes on the cot graph are at x=πk; thus the domain is all real numbers except for these asymptotes.

Notice that there is no *y* intercept, and the *x* intercepts are multiples of $\frac{\pi}{2}$; we can write this as $\left(\frac{\pi}{2}+\pi k,0\right)$ where $k \in I$ (*k* is in the set of Integers).

Note that the cot function is an odd function; Thus the cot graph is symmetrical about the origin (0, 0), meaning that if (*x*, *y*) is a point on the function (graph), then so is (−*x*, −*y*). It also means that for the cot graph, $f(-x)=-f(x)$. let's try this: $\cot\left(-\frac{\pi}{4}\right)=-1=-\cot\left(\frac{\pi}{4}\right)$.

Graph of Inverse Trigonometric Functions

The inverse sine function, denoted by $\sin^{-1} x$ (or arcsin x), is defined to be the inverse of the restricted sine function

- $\sin x, \quad -\frac{\pi}{2} \le x \le \frac{\pi}{2}$

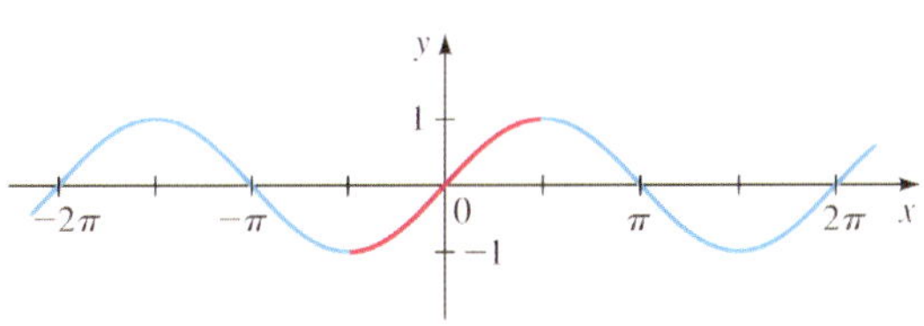

$y = \sin x$

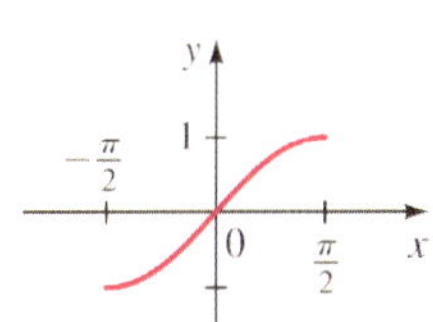

$y = \sin x,\ -\frac{\pi}{2} \le x \le \frac{\pi}{2}$

−1

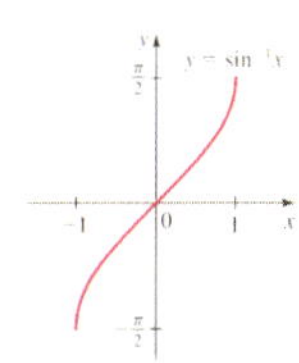

The inverse cosine function, denoted by $\cos^{-1} x$ (or arccos x), is defined to be the inverse of the restricted cosine function

$$\cos x,\ 0 \le x \le \pi$$

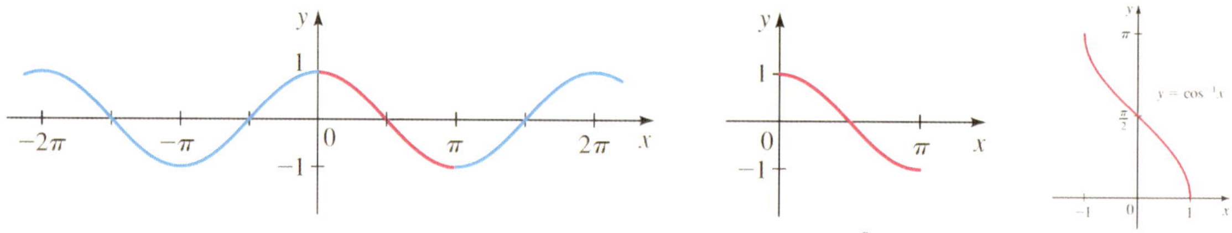

$y = \cos x$ $y = \cos x,\ 0 \le x \le \pi$ $y = \cos^{-1} x$

The inverse tangent function, denoted by $\tan^{-1} x$ (or arctan x), is defined to be the inverse of the restricted tangent function

$$\tan x,\quad -\frac{\pi}{2} < x < \frac{\pi}{2}$$

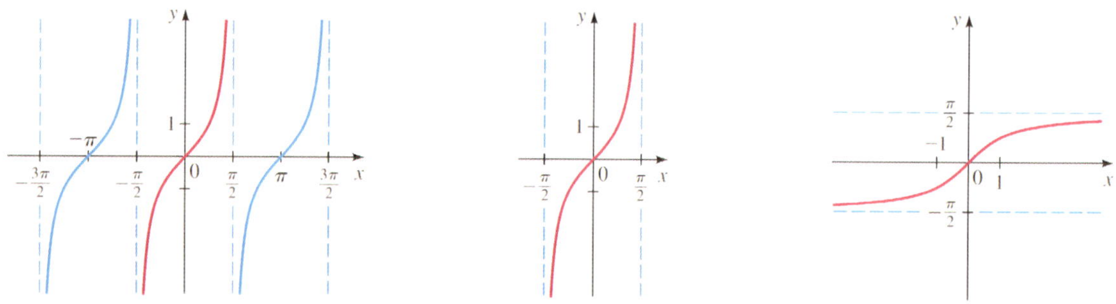

$y = \tan x$ $y = \tan x,\ -\frac{\pi}{2} < x < \frac{\pi}{2}$ $y = \tan^{-1} x$

The inverse cotangent function, denoted by $\cot^{-1} x$ (or arccot x), is defined to be the inverse of the restricted cotangent function

$$\cot x,\ 0 < x < \pi$$

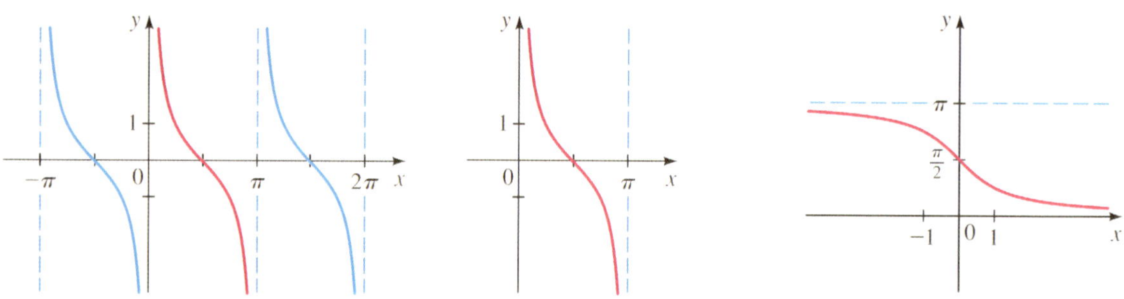

$y = \cot x$ $y = \cot x,\ 0 < x < \pi$ $y = \cot^{-1} x$

The inverse secant function, denoted by $\sec^{-1} x$ (or arcsec x), is defined to be the inverse of the restricted secant function

$$\sec x,\quad x \in [0, \pi/2) \cup [\pi, 3\pi/2) \quad \left[\text{or } x \in [0, \pi/2) \cup (\pi/2, \pi] \text{ in some other text books}\right]$$

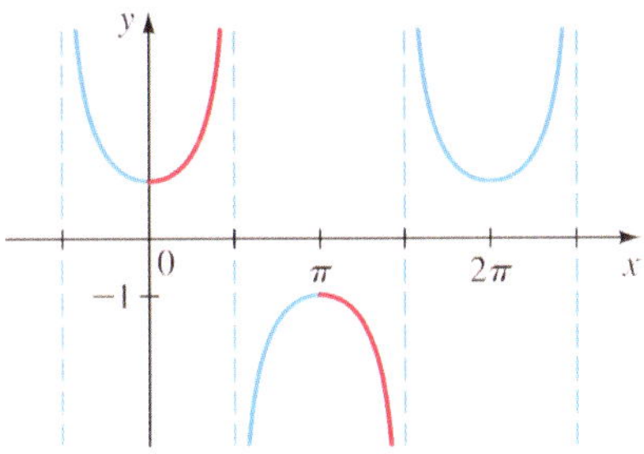

$y = \sec x$

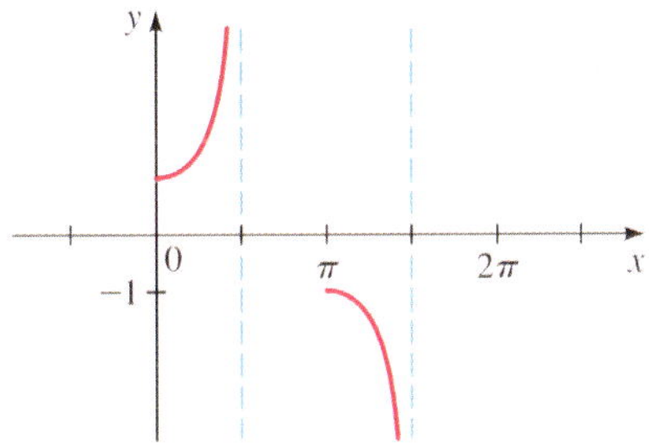

$y = \sec x,\ 0 \le x < \frac{\pi}{2},\ \pi \le x < \frac{3\pi}{2}$

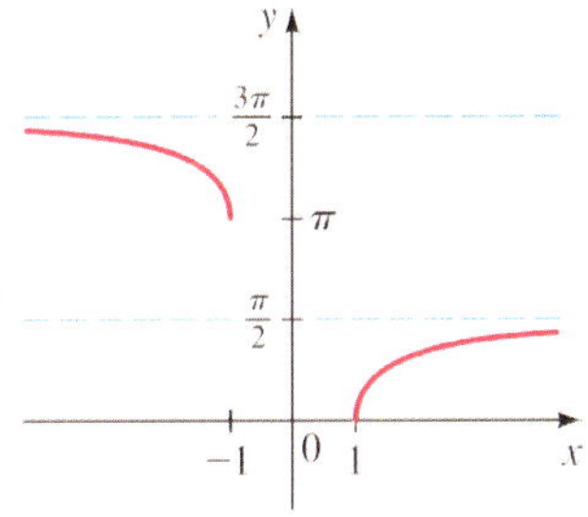

$y = \sec^{-1} x$

The inverse cosecant function, denoted by $\csc^{-1} x$ (or arccsc x), is defined to be the inverse of the restricted cosecant function

$$\csc x, \quad x \in [0, \pi/2) \cup [\pi, 3\pi/2) \quad [\text{or } x \in [-\pi/2, 0) \cup (0, \pi/2] \text{ in some other text books}]$$

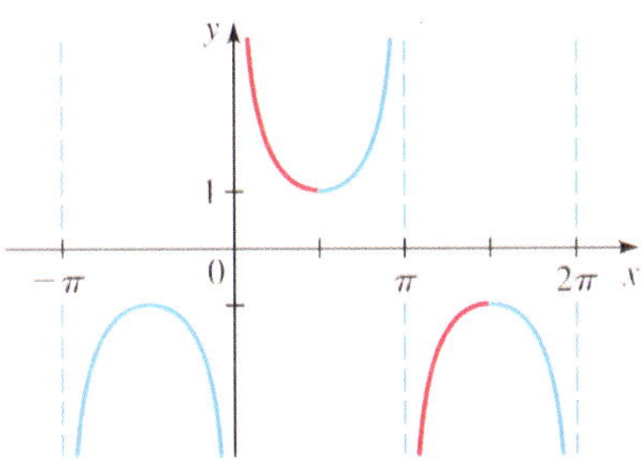

$y = \csc x$

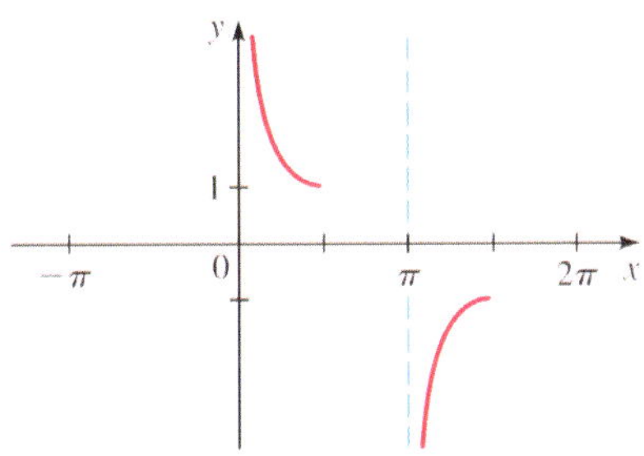

$y = \csc x,\ 0 < x \le \frac{\pi}{2},\ \pi < x \le \frac{3\pi}{2}$

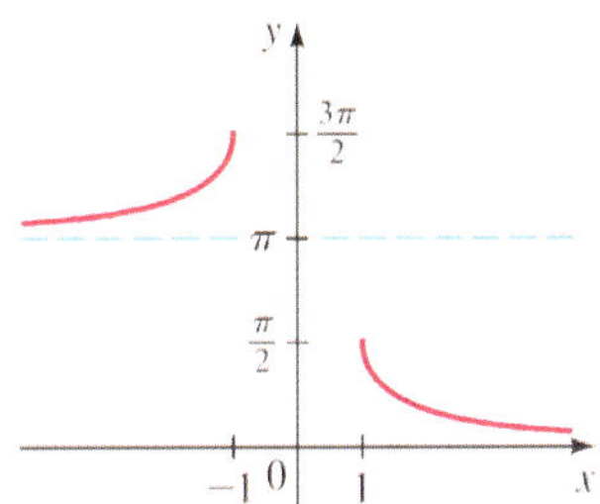

$y = \csc^{-1} x$

References

- Graphs-sine-cosine-tangent, trigonometry: shelovesmath.com, Retrieved 29 May 2018
- Period-of-a-function, calculus: math.tutorvista.com, Retrieved 31 March 2018
- Graph-of-y-equals-tan-x: math-only-math.com, Retrieved 16 April 2018
- 4-graphs-tangent-cotangent-secant-cosecant, trigonometric-graphs: intmath.com, Retrieved 21 May 2018
- Graphs-sine-cosine-tangent, trigonometry: shelovesmath.com, Retrieved 25 June 2018

4

Laws and Theorems in Trigonometry

Trigonometry is an important area of mathematics that is governed by a number of laws and theorems. This chapter closely examines the fundamental laws and theorems in trigonometry to provide an extensive understanding of the subject. It includes topics like law of sines, law of cosines, law of tangents, Pythagorean theorem, Thales's theorem, etc.

Law of Sines

The Law of Sine is the relationship between the sides and angles of non-right (oblique) triangles . Simply, it states that the ratio of the length of a side of a triangle to the sine of the angle opposite that side is the same for all sides and angles in a given triangle.

In ΔABC is an oblique triangle with sides a, b and c, then $\frac{a}{\sin A} = \frac{b}{\sin B} = \frac{c}{\sin C}$

Oblique Triangles

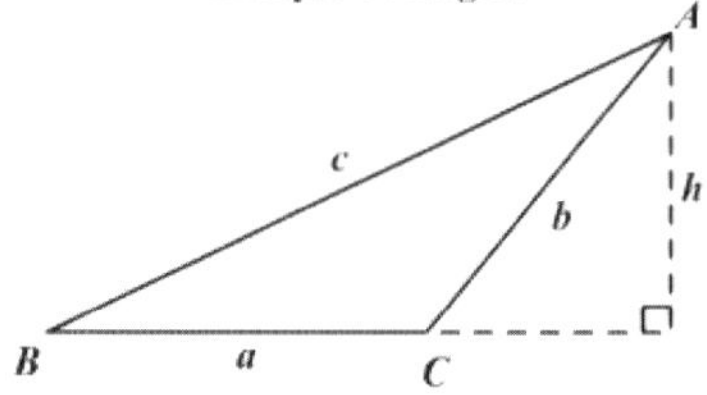

C is Obtuse

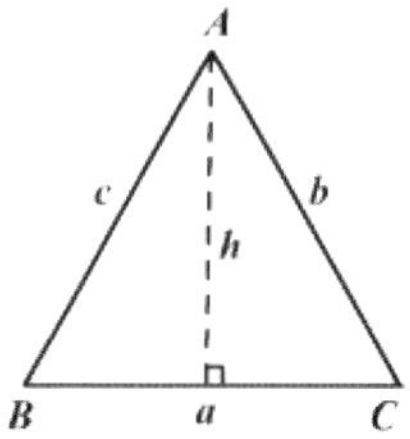

C is Acute

To use the Law of Sine you need to know either two angles and one side of the triangle (AAS or ASA) or two sides and an angle opposite one of them (SSA). Notice that for the first two cases we use the same parts that we used to prove congruence of triangles in geometry but in the last case we could not prove congruent triangles given these parts. This is because the remaining pieces could have been different sizes. This is called the ambiguous case and we will discuss it a little later.

Example 1: Given two angles and a non-included side (AAS)

Given ΔABC with $m\angle A = 30°$, $m\angle B = 20°$ and $a = 45$ m. Find the remaining angle and sides.

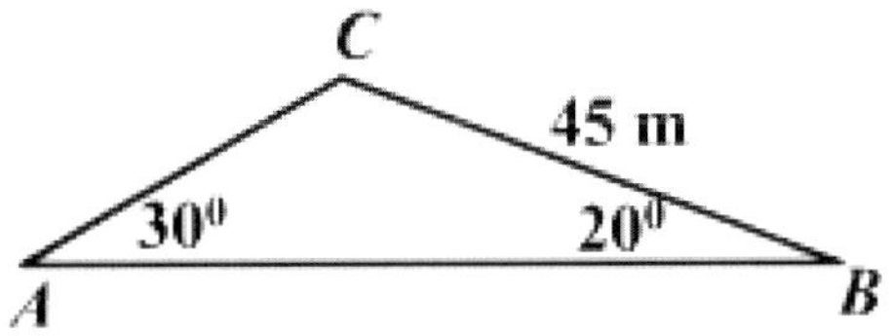

The third angle of the triangle is

$$m\angle C = 180° - m\angle A - m\angle B = 180° - 30° - 20° = 130°$$

By the Law of Sine,

$$\frac{45}{\sin 30^{\circ}} = \frac{b}{\sin 20^{\circ}} = \frac{c}{\sin 130^{\circ}}$$

By the Properties of Proportions

$$b = \frac{45 \sin 20^{\circ}}{\sin 30^{\circ}} \approx 30.78 \text{ m and } c = \frac{45 \sin 130^{\circ}}{\sin 30^{\circ}} \approx 68.94 \text{m}$$

Example 2: Given two angles and an included side (ASA).

Given $m\angle A = 42°$, $m\angle B = 75°$ and $c = 22\text{cm}$. Find the remaining angle and sides.

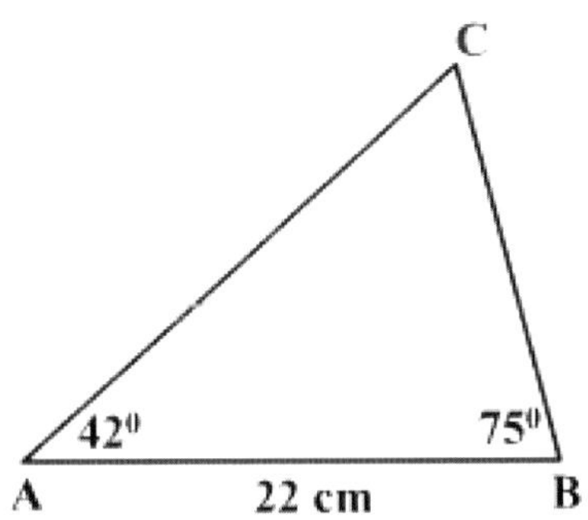

The third angle of the triangle is:

$$m\angle C = 180° - m\angle A - m\angle B = 180° - 42° - 75° = 63°$$

By thc Law of Sinc,

$$\frac{a}{\sin 42^{\circ}} = \frac{b}{\sin 75^{\circ}} = \frac{c}{\sin 63^{\circ}}$$

By the Properties of Proportions

$$a = \frac{22 \sin 42^{\circ}}{\sin 63^{\circ}} \approx 16.52 \text{ cm and } b = \frac{22 \sin 75^{\circ}}{\sin 63^{\circ}} \approx 23.85 \text{cm}$$

The Ambiguous Case

If two sides and an angle opposite one of them are given, three possibilities can occur.

(1) No such triangle exists.

(2) Two different triangles exist.

(3) Exactly one triangle exists.

Consider a triangle in which you are given a, b and A. (The altitude h from vertex B to side $\overline{AC}$, by the definition of sines is equal to b sin A.)

(1) No such triangle exists if A is acute and $a < h$ or A is obtuse and $a \le b$.

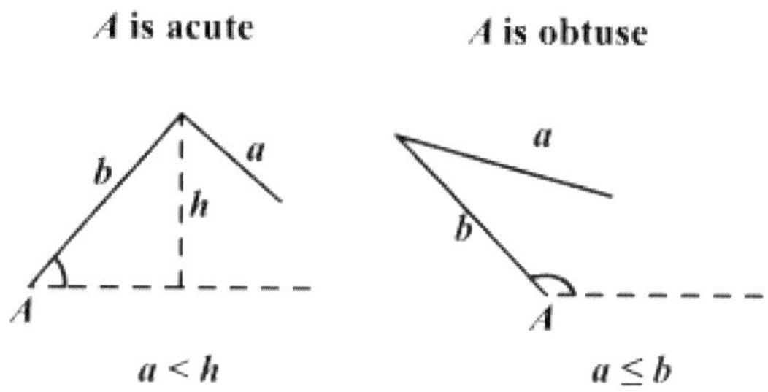

2) Two different triangles exist if A is acute and $h < a < b$.

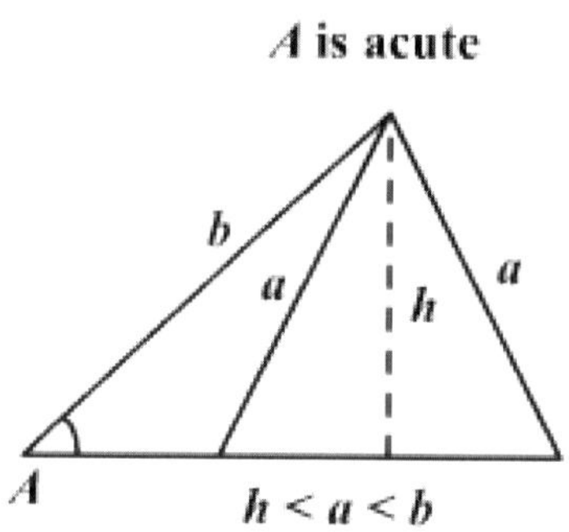

(3) In every other case, exactly one triangle exists.

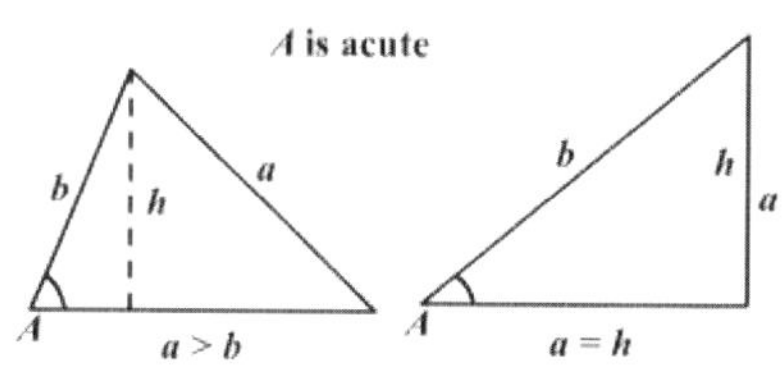

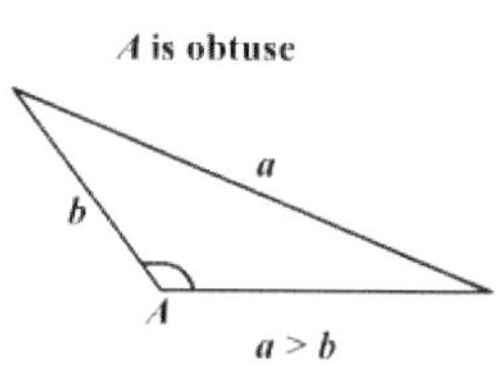

Example: No Solution Exists

Given $a = 15, b = 25$ and $m\angle A = 80°$. Find the other angles and side.

$$h = b\sin A = 25\sin 80° \approx 24.6$$

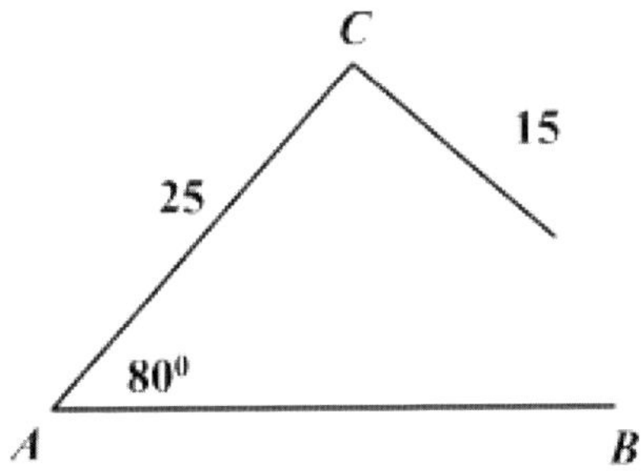

Notice that $a < h$ so it appears that there is no solution. Verify this using the Law of Sine.

$$\frac{a}{\sin A} = \frac{b}{\sin B}$$

$$\frac{15}{\sin 80°} = \frac{25}{\sin B}$$

$$\sin B = \frac{25\sin 80°}{15} \approx 1.641 > 1$$

This contrasts the fact that the $-1 \le \sin B \le 1$.Therefore, no triangle exists.

Example: Two Solutions Exist

Given $a = 6, b = 7$ and $m\angle A = 30°$. Find the other angles and side.

$$h = b\sin A = 7\sin 30° = 3.5$$

$h < a < b$ Therefore, there are two triangles possible.

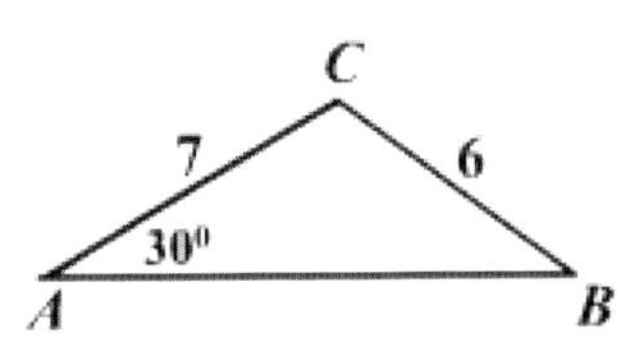

$\angle$ ***B* is acute**

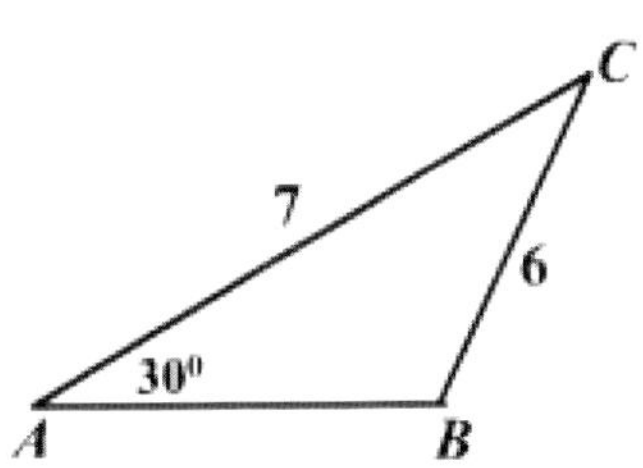

$\angle$ ***B* is obtuse**

By the Law of Sine, $\dfrac{a}{\sin A}=\dfrac{b}{\sin B}$

$$\sin B=\frac{b\sin A}{a}=\frac{7\sin 30^{\circ}}{6}\approx 0.5833$$

There are two angles between 0°and 180°whose sine is approximately 0.5833, are 35.69°and 144.31°.

If $B\approx 35.69^{\circ}$

$C\approx 180^{\circ}-30^{\circ}-35.69^{\circ}=114.31^{\circ}$

$$c=\frac{a\sin C}{\sin A}\approx\frac{6\sin 114.31^{\circ}}{\sin 30^{\circ}}\approx 10.94$$

If $B\approx 144.31^{\circ}$

$C\approx 180^{\circ}-30^{\circ}-144.31^{\circ}=5.69^{\circ}$

$$c\approx\frac{6\sin 5.69^{\circ}}{\sin 30^{\circ}}\approx 1.19$$

Example: One Solution Exists

Given $a=22$, $b=12$ and $m\angle A=40^{\circ}$. Find the other angles and side.

$a>b$

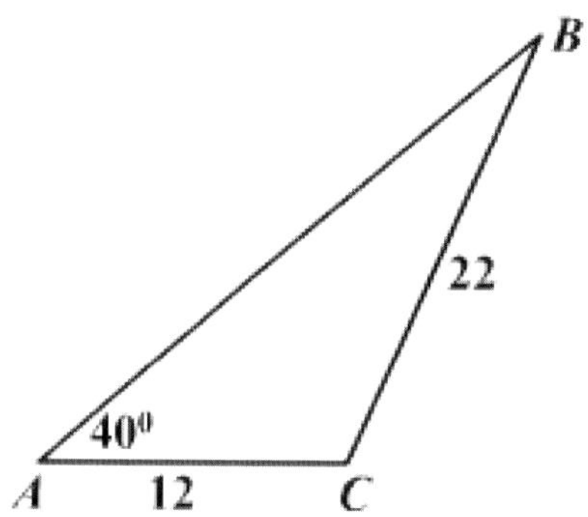

By the Law of Sine, $\dfrac{a}{\sin A}=\dfrac{b}{\sin B}$

$$\sin B=\frac{b\sin A}{a}=\frac{12\sin 40^{\circ}}{22}\approx 0.3506$$

$B\approx 20.52^{\circ}$

B is acute.

$$m\angle C=180^{\circ}-m\angle A-m\angle B=180^{\circ}-40^{\circ}-20.52^{\circ}=29.79^{\circ}$$

By the Law of Sine,

$$\frac{c}{\sin 119.48^{\circ}}=\frac{22}{\sin 40^{\circ}}$$

$$c=\frac{22\sin 119.48^{\circ}}{\sin 40^{\circ}}\approx 29.79$$

If we are given two sides and an included angle of a triangle or if we are given 3 sides of a triangle, we cannot use the Law of Sine because we cannot set up any proportions where enough information is known.

Proof

The area T of any triangle can be written as one half of its base times its height. Selecting one side of the triangle as the base, the height of the triangle relative to that base is computed as the length of another side times the sine of the angle between the chosen side and the base. Thus, depending on the selection of the base the area of the triangle can be written as any of:

$$T = \frac{1}{2}b(c\sin A) = \frac{1}{2}c(a\sin B) = \frac{1}{2}a(b\sin C).$$

Multiplying these by $\frac{2}{abc}$ gives

$$\frac{2T}{abc} = \frac{\sin A}{a} = \frac{\sin B}{b} = \frac{\sin C}{c}.$$

Relation to the Circumcircle

In the identity

$$\frac{a}{\sin A} = \frac{b}{\sin B} = \frac{c}{\sin C},$$

The common value of the three fractions is actually the diameter of the triangle's circumcircle which dates back to Ptolemy.

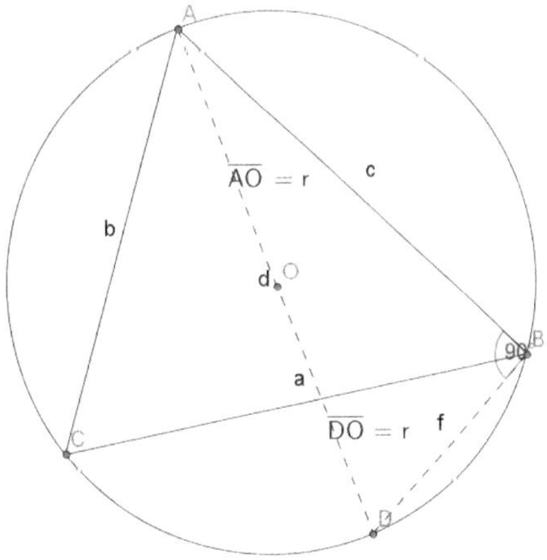

Deriving the ratio of the sine law equal to the circumscribing diameter. Note that triangle ADB passes through the center of the circumscribing circle.

Proof

As shown in the figure, let's have a circle with inscribed ΔABC and another inscribed ΔADB that passes through the circle's center O. The $\angle AOD$ has a central angle of 180° thus $\angle ABD = 90^\circ$. Since ΔABD is a right triangle,

$$\sin D = \frac{opposite}{hypotenuse} = \frac{c}{2r}$$

Angles $\angle C$ and $\angle D$ has the same central angle thus they are the same $\angle C = \angle D$.

$$\sin D = \sin C = \frac{c}{2r}$$

Rearranging,

$$2r = \frac{c}{\sin C}$$

Repeating the process of creating ΔADB with other points respectively, gives

$$\frac{a}{\sin A} = \frac{b}{\sin B} = \frac{c}{\sin C} = 2r$$

Where r is the radius of the circumscribing circle of the triangle.

Relationship to the Area of the Triangle

The area of a triangle can be solved by $A = 0.5ab\sin\theta$.. Substituting the sine law to the equation,

$$A = \frac{1}{2}ab \cdot \frac{c}{2r}$$

Taking r as the circumscribing radius,

$$A = \frac{abc}{4r}$$

It can also be shown that this quantity is equal to

$$\frac{abc}{2T} = \frac{abc}{2\sqrt{s(s-a)(s-b)(s-c)}}$$

$$= \frac{2abc}{\sqrt{(a^2+b^2+c^2)^2 - 2(a^4+b^4+c^4)}},$$

Where T is the area of the triangle and s is the semiperimeter

$$s = \frac{a+b+c}{2}.$$

The second equality above readily simplifies to Heron's formula for the area. Neatly, it can be written as

$$r = \frac{abc}{4T}$$

where r is the radius of the triangle's circumcircle.

Curvature

The law of sines takes on a similar form in the presence of curvature.

Spherical Case

In the spherical case, the formula is:

$$\frac{\sin A}{\sin \alpha} = \frac{\sin B}{\sin \beta} = \frac{\sin C}{\sin \gamma}.$$

Here, α, β, and γ are the angles at the center of the sphere subtended by the three arcs of the spherical surface triangle a, b, and c, respectively. A, B, and C are the surface angles opposite their respective arcs.

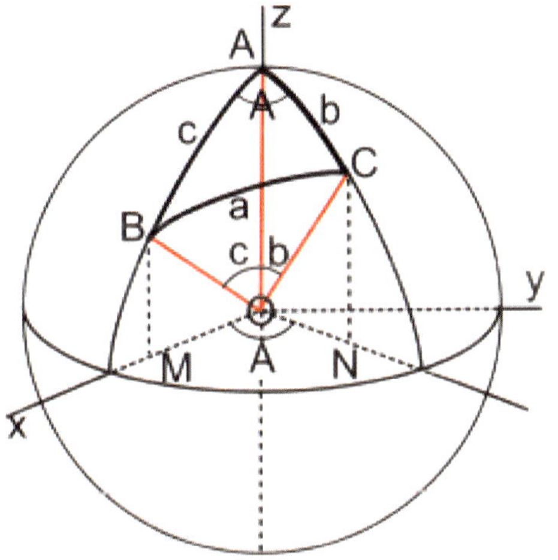

Here is a simple derivation. Consider a unit sphere with three unit vectors OA, OB and OC drawn from the origin to the vertices of the triangle. Thus the angles α, β, and γ are the angles a, b, and c, respectively. The arc BC subtends an angle of magnitude a at the center. Introduce a Cartesian basis with OA along the z-axis and OB in the xz-plane making an angle c with the z-axis. The vector OC projects to ON in the xy-plane and the angle between ON and the x-axis is A. Therefore, the three vectors have components:

$$\mathbf{OA} = \begin{pmatrix} 0 \\ 0 \\ 1 \end{pmatrix}, \quad \mathbf{OB} = \begin{pmatrix} \sin c \\ 0 \\ \cos c \end{pmatrix}, \quad \mathbf{OC} = \begin{pmatrix} \sin b \cos A \\ \sin b \sin A \\ \cos b \end{pmatrix}.$$

The scalar triple product, OA · (OB × OC) is the volume of the parallelepiped formed by the position vectors of the vertices of the spherical triangle OA, OB and OC. This volume is invariant to the specific coordinate system used to represent OA, OB and OC. The value of the scalar triple product OA · (OB × OC) is the 3 × 3 determinant with OA, OB and OC as its rows. With the z-axis along OA the square of this determinant is

$$(\mathrm{OA}\cdot(\mathrm{OB}\times\mathrm{OC}))^2 = (\det(\mathrm{OA},\mathrm{OB},\mathrm{OC}))^2$$

$$= \left(\begin{vmatrix} 0 & 0 & 1 \\ \sin c & 0 & \cos c \\ \sin b\cos A & \sin b\sin A & \cos b \end{vmatrix}\right)^2 = (\sin b\sin c\sin A)^2.$$

Repeating this calculation with the z-axis along OB gives $(\sin c \sin a \sin B)^2$, while with the z-axis along OC it is $(\sin a \sin b \sin C)^2$. Equating these expressions and dividing throughout by $(\sin a \sin b \sin c)^2$ gives

$$\frac{\sin^2 A}{\sin^2 a} = \frac{\sin^2 B}{\sin^2 b} = \frac{\sin^2 C}{\sin^2 c} = \frac{V^2}{\sin^2 a \sin^2 b \sin^2 c},$$

where V is the volume of the parallelepiped formed by the position vector of the vertices of the spherical triangle.

It is easy to see how for small spherical triangles, when the radius of the sphere is much greater than the sides of the triangle, this formula becomes the planar formula at the limit, since

$$\lim_{\alpha\to 0}\frac{\sin\alpha}{\alpha} = 1$$

and the same for $\sin\beta$ and $\sin\gamma$.

Hyperbolic Case

In hyperbolic geometry when the curvature is −1, the law of sines becomes

$$\frac{\sin A}{\sinh a} = \frac{\sin B}{\sinh b} = \frac{\sin C}{\sinh c}.$$

In the special case when B is a right angle, one gets

$$\sin C = \frac{\sinh c}{\sinh b}$$

which is the analog of the formula in Euclidean geometry expressing the sine of an angle as the opposite side divided by the hypotenuse.

Unified Formulation

Define a generalized sine function, depending also on a real parameter K:

$$\sin_K x = x - \frac{Kx^3}{3!} + \frac{K^2x^5}{5!} - \frac{K^3x^7}{7!} + \cdots.$$

The law of sine in constant curvature K reads as

$$\frac{\sin A}{\sin_K a} = \frac{\sin B}{\sin_K b} = \frac{\sin C}{\sin_K c}.$$

By substituting $K = 0$, $K = 1$, and $K = -1$, one obtains respectively the Euclidean, spherical, and hyperbolic cases of the law of sine described above.

Let $p_K(r)$ indicate the circumference of a circle of radius r in a space of constant curvature K. Then $p_K(r) = 2\pi \sin_K r$. Therefore the law of sine can also be expressed as:

$$\frac{\sin A}{p_K(a)} = \frac{\sin B}{p_K(b)} = \frac{\sin C}{p_K(c)}.$$

This formulation was discovered by János Bolyai.

Higher Dimensions

For an n-dimensional simplex (i.e., triangle ($n = 2$), tetrahedron ($n = 3$), pentatope ($n = 4$), etc.) in n-dimensional Euclidean space, the absolute value of the polar sine of the normal vectors of the faces that meet at a vertex, divided by the hyperarea of the face opposite the vertex is independent of the choice of the vertex. Writing V for the hyper volume of the n-dimensional simplex and P for the product of the hyperareas of its $(n-1)$-dimensional faces, the common ratio is

$$\frac{(nV)^{n-1}}{(n-1)!P}.$$

For example, a tetrahedron has four triangular faces. The absolute value of the polar sine of the normal vectors to the three faces that share a vertex, divided by the area of the fourth face will not depend upon the choice of the vertex:

$$\frac{|\operatorname{psin}(\mathbf{n}_2,\mathbf{n}_3,\mathbf{n}_4)|}{\text{Area}_1} = \frac{|\operatorname{psin}(\mathbf{n}_1,\mathbf{n}_3,\mathbf{n}_4)|}{\text{Area}_2} = \frac{|\operatorname{psin}(\mathbf{n}_1,\mathbf{n}_2,\mathbf{n}_4)|}{\text{Area}_3} = \frac{|\operatorname{psin}(\mathbf{n}_1,\mathbf{n}_2,\mathbf{n}_3)|}{\text{Area}_4}$$

$$= \frac{(3\,\text{Volume}_{\text{tetrahedron}})^2}{2!\,\text{Area}_1\text{Area}_2\text{Area}_3\text{Area}_4}.$$

Law of Cosines

This law basically provides us with a mathematical equation that relates the three sides of a triangle and one of its angles. it is no different from the Pythagorean theorem. In fact, it is a little bit more general than the Pythagorean Theorem in the sense that it applies to basically any kind of triangle (right triangle or some odd triangle).

Consider the following picture of a *general* triangle with the sides' labeled as *a, b, and c*:

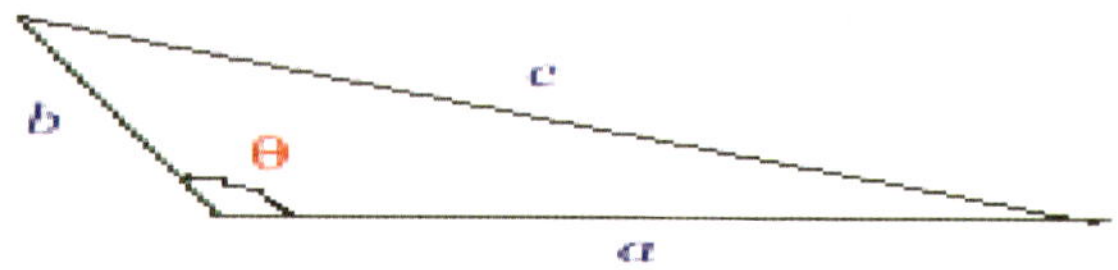

Figure: Three sides and the angle θ related by the law of cosiness,
Note that θ is the angle opposite to the side c.

The law of cosines for the above triangle relates the three sides to the angle that we have labeled as θ as follows:

$$c^2 = a^2 + b^2 - 2a.b.\cos(\theta)$$

The angle θ is the angle that is *opposite* to the side labeled *c*.

There are two other versions of the law of cosines,

$$a^2 = b^2 + c^2 - 2bc\cos A$$

and

$$b^2 = a^2 + c^2 - 2ac\cos B.$$

Since the three verions differ only in the labelling of the triangle, it is enough to verify one just one of them. We'll consider the version stated first.

In order to see why these laws are valid, we'll have to look at three cases. For case 1, we'll take the angle *C* to be obtuse. In case 2, angle *C* will be a right angle. In case 3, angle *C* will be acute.

Case 1: For this case, we take angle C to be obtuse. This case has a wrinkle in it since the cosine of an obtuse angle is negative.

First, drop a perpendicular line AD from A down to the base BC of the triangle. In this case, the foot D of this perpendicular will lie outside the triangle. Let h denote the height of the triangle, let *d* denote BD, and let e denote CD.

We can derive the following equations from the figure:

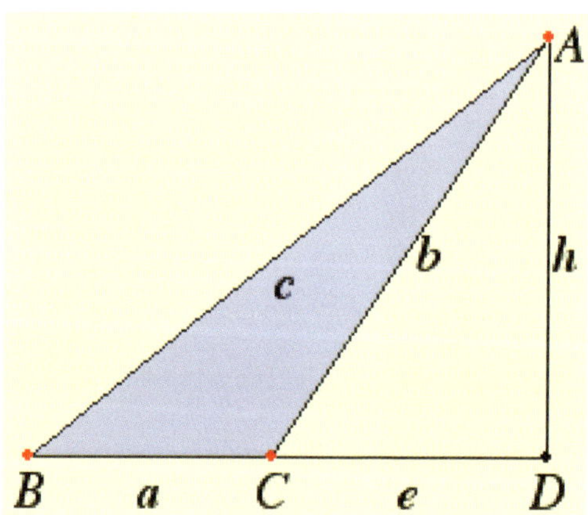

$$\begin{aligned} c^2 &= d^2 + h^2 \\ b^2 &= e^2 + h^2 \\ d &= a + e \\ \cos\theta &= -e/b \end{aligned}$$

In general, the cosine of an obtuse angle is the negation of the cosine of its supplement. In this case that means the cosine of angle C, that is to say angle ACB, is the negation of the cosine of angle ACD. That's why the minus sign appears in the last equation.

These equations and plain algebra finish the argument as follows:

$$\begin{aligned} c^2 &= d^2 + h^2 \\ &= (a+e)^2 + h^2 \\ &= a^2 + 2ae + e^2 + h^2 \\ &= a^2 + b^2 + 2ae \\ &= a^2 + b^2 - 2ab\cos\theta \end{aligned}$$

Thus, the law of cosines is valid when *C* is an obtuse angle.

Case 2: Now consider the case when the angle at *C* is right. The cosine of a right angle is 0, so the law of cosines, $c^2 = a^2 + b^2 - 2ab\ \cos\theta$, simplifies to becomes the Pythagorean identity, $c^2 = a^2 + b^2$, for right triangles which we know is valid.

Case 3: In this case we assume that the angle *C* is an acute triangle. Drop a perpendicular line *AD* from *A* down to the base *BC* of the triangle. The foot *D* of the perpendicular will (1) lie on the edge *BC* if angle *B* is acute, (2) coincide with the point *B* if the angle *B* is right, or (3) lie on the side *BC* extended if the angle *B* is obtuse.

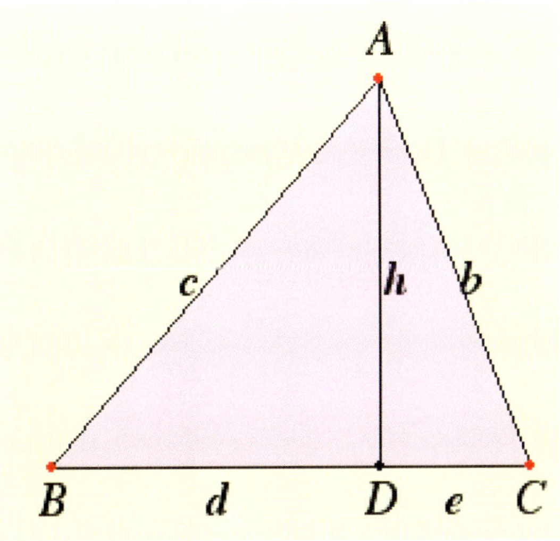

Let *h* denote the height of the triangle, let *d* denotes *BD*, and *e* denote *CD*.

Then we can read the following relationships from the diagram:

$$C^2 = d^2 + h^2$$
$$b^2 = e^2 + h^2$$
$$\cos\theta = e/b$$
$$d^2 = (e-a)^2$$

That last equation requires explanation. If the point *D* lies on the side *BC*, then $d = a - e$, but if *D* lies on *BC* extended, then $d = e - a$. In either case, $d^2 = (e - a)^2$.

These equations and a little algebra finish the proof as follows:

Thus, we now know that the law of cosines is valid when both angle *C* is acute,

$$\begin{aligned} C^2 &= d^2 + h^2 \\ &= d^2 - e^2 + b^2 \\ &= (d-e)(d+e) + b^2 \\ &= (a-2e)a + b^2 \\ &= a^2 + b^2 - 2ae \\ &= a^2 + b^2 - 2ab\cos\theta \end{aligned}$$

Thus, we now know that the law of cosines is valid when both angle C is acute.

Proofs

Using the Distance Formula

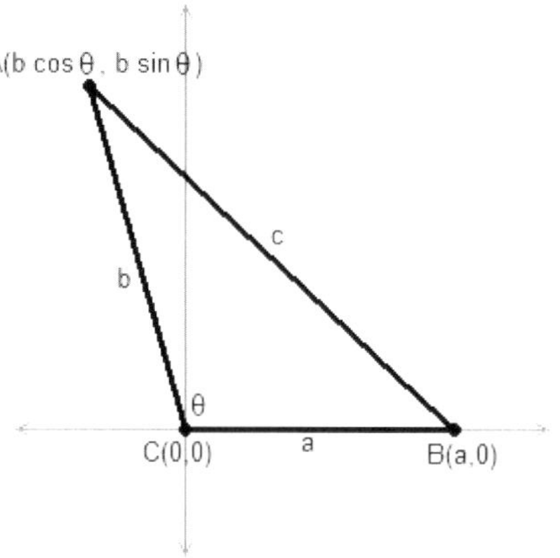

Figure: Coordinate geometry proof

Consider a triangle with sides of length *a*, *b*, *c*, where θ is the measurement of the angle opposite the side of length *c*. This triangle can be placed on the Cartesian coordinate system by plotting the following points, as shown in above Figure:

$$A = (b\cos\theta, b\sin\theta),\ B = (a, 0),\ \text{and}\ C = (0, 0).$$

By the distance formula, we have

$$c = \sqrt{(a - b\cos\theta)^2 + (0 - b\sin\theta)^2}.$$

Now, we just work with that equation:

$$c^2 = (a - b\cos\theta)^2 + (-b\sin\theta)^2$$
$$c^2 = a^2 - 2ab\cos\theta + b^2\cos^2\theta + b^2\sin^2\theta$$
$$c^2 = a^2 + b^2(\sin^2\theta + \cos^2\theta) - 2ab\cos\theta$$
$$c^2 = a^2 + b^2 - 2ab\cos\theta.$$

An advantage of this proof is that it does not require the consideration of different cases for when the triangle is acute vs. right vs. obtuse.

Using Trigonometry

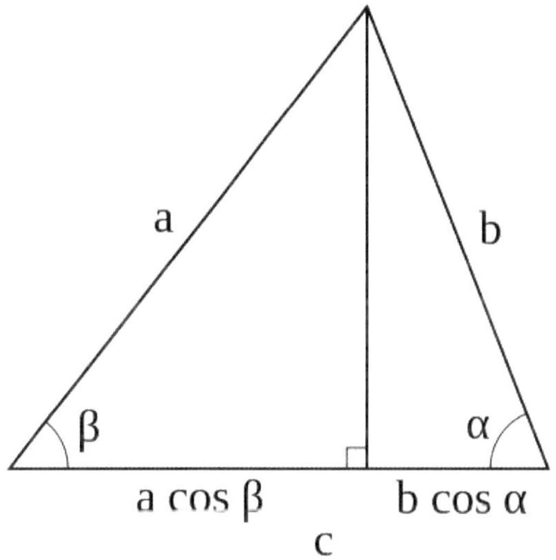

Figure: An acute triangle with perpendicular

Drop the perpendicular onto the side c to get (see Figure)

$$c = a\cos\beta + b\cos\alpha.$$

(This is still true if α or β is obtuse, in which case the perpendicular falls outside the triangle.) Multiply through by c to get

$$c^2 = ac\cos\beta + bc\cos\alpha.$$

By considering the other perpendiculars obtain

$$a^2 = ac\cos\beta + ab\cos\gamma,$$

$$b^2 = bc\cos\alpha + ab\cos\gamma.$$

Adding the latter two equations gives

$$a^2 + b^2 = ac\cos\beta + bc\cos\alpha + 2ab\cos\gamma.$$

Subtracting the first equation from the last one we have

$$a^2 + b^2 - c^2 = -ac\cos\beta - bc\cos\alpha + ac\cos\beta + bc\cos\alpha + 2ab\cos\gamma$$

which simplifies to

$$c^2 = a^2 + b^2 - 2ab\cos\gamma.$$

This proof uses trigonometry in that it treats the cosines of the various angles as quantities in their own right. It uses the fact that the cosine of an angle expresses the relation between the two sides enclosing that angle in *any* right triangle. Other proofs (below) are more geometric in that they treat an expression such as $a \cos \gamma$ merely as a label for the length of a certain line segment.

Many proofs deal with the cases of obtuse and acute angles γ separately.

Using Ptolemy's Theorem

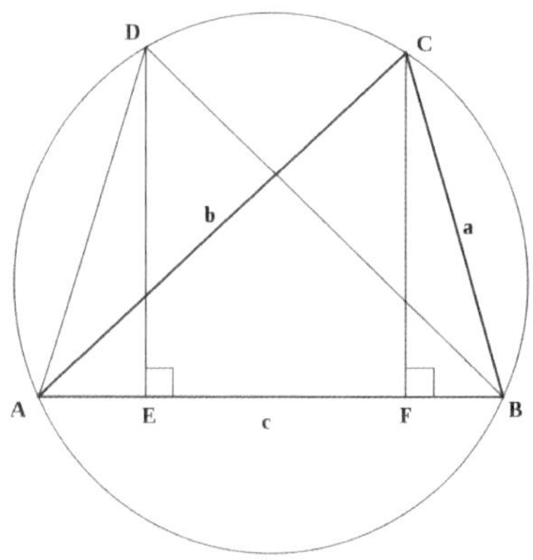

Proof of law of cosines using Ptolemy's theorem

Referring to the diagram, triangle ABC with sides $AB = c$, $BC = a$ and $AC = b$ is drawn inside its circumcircle as shown. Triangle ABD is constructed congruent to triangle ABC with $AD = BC$ and $BD = AC$. Perpendiculars from D and C meet base AB at E and F respectively. Then:

$$BF = AE = BC\cos\hat{B} = a\cos\hat{B}$$
$$\Rightarrow DC = EF = AB - 2BF = c - 2a\cos\hat{B}.$$

Now the law of cosines is rendered by a straightforward application of Ptolemy's theorem to cyclic quadrilateral $ABCD$:

$$AD \times BC + AB \times DC = AC \times BD$$
$$\Rightarrow a^2 + c(c - 2a\cos\hat{B}) = b^2$$
$$\Rightarrow a^2 + c^2 - 2ac\cos\hat{B} = b^2.$$

Plainly if angle B is right, then $ABCD$ is a rectangle and application of Ptolemy's theorem yields the Pythagorean theorem:

$$a^2 + c^2 = b^2.$$

By Comparing Areas

One can also prove the law of cosines by calculating areas. The change of sign as the angle γ becomes obtuse makes a case distinction necessary.

Recall that

- a^2, b^2, and c^2 are the areas of the squares with sides a, b, and c, respectively;
- if γ is acute, then $ab \cos \gamma$ is the area of the parallelogram with sides a and b forming an angle of $\gamma' = \frac{\pi}{2} - \gamma$;
- if γ is obtuse, and so $\cos \gamma$ is negative, then $-ab \cos \gamma$ is the area of the parallelogram with sides a and b forming an angle of $\gamma' = \gamma - \frac{\pi}{2}$.

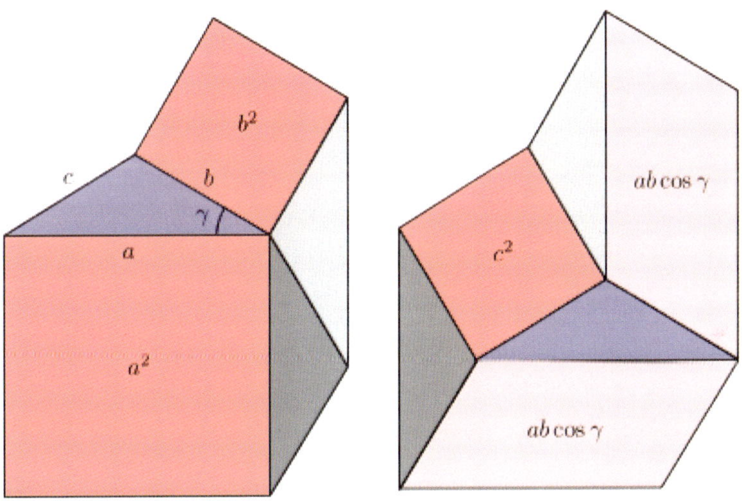

Figure: Proof of the law of cosines for acute angle γ by "cutting and pasting".

Acute case Figure shows a heptagon cut into smaller pieces (in two different ways) to yield a proof of the law of cosines. The various pieces are

- in pink, the areas a^2, b^2 on the left and the areas $2ab \cos \gamma$ and c^2 on the right;
- in blue, the triangle ABC, on the left and on the right;
- in grey, auxiliary triangles, all congruent to ABC, an equal number (namely 2) both on the left and on the right.

The equality of areas on the left and on the right gives

$$a^2 + b^2 = c^2 + 2ab \cos \gamma.$$

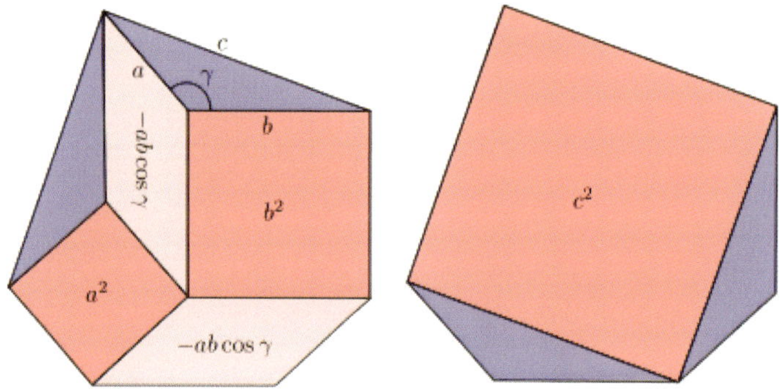

Figure: Proof of the law of cosines for obtuse angle γ by "cutting and pasting".

Obtuse case Figure cuts a hexagon in two different ways into smaller pieces, yielding a proof of the law of cosines in the case that the angle γ is obtuse. We have

- in pink, the areas a^2, b^2, and $-2ab\cos\gamma$ on the left and c^2 on the right;
- in blue, the triangle ABC twice, on the left, as well as on the right.

The equality of areas on the left and on the right gives

$$a^2 + b^2 - 2ab\cos(\gamma) = c^2.$$

The rigorous proof will have to include proofs that various shapes are congruent and therefore have equal area. This will use the theory of congruent triangles.

Using geometry of the Circle

Using the geometry of the circle, it is possible to give a more geometric proof than using the Pythagorean theorem alone. Algebraic manipulations (in particular the binomial theorem) are avoided.

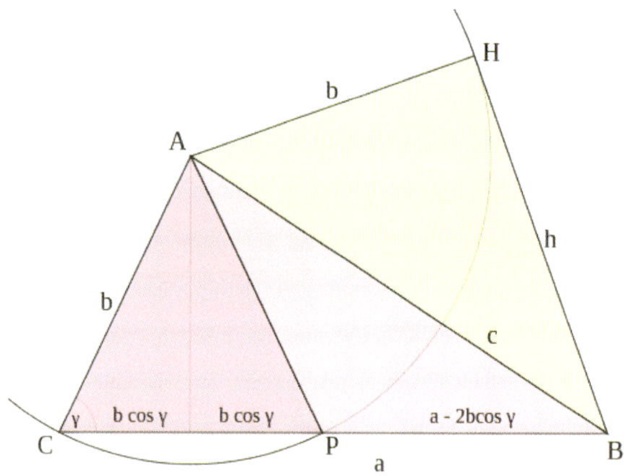

Figure: The triangle ABC (pink), an auxiliary circle (light blue) and an auxiliary right triangle (yellow)

Case of acute angle γ, where $a > 2b\cos\gamma$. Drop the perpendicular from A onto $a = BC$, creating a line segment of length $b\cos\gamma$. Duplicate the *right triangle* to form the isosceles triangle ACP. Construct the circle with center A and radius b, and its tangent $h = BH$ through B. The tangent h forms a right angle with the radius b, so the yellow triangle in Figure is right. Apply the Pythagorean theorem to obtain,

$$c^2 = b^2 + h^2.$$

Then use the *tangent secant theorem* (Euclid's *Elements*: Book 3, Proposition 36), which says that the square on the tangent through a point B outside the circle is equal to the product of the two lines segments (from B) created by any secant of the circle through B. In the present case: $BH^2 = BC \cdot BP$, or

$$h^2 = a(a - 2b\cos\gamma).$$

Substituting into the previous equation gives the law of cosines:

$$c^2 = b^2 + a(a - 2b\cos\gamma).$$

Note that h^2 is the power of the point B with respect to the circle. The use of the Pythagorean theorem and the tangent secant theorem can be replaced by a single application of the power of a point theorem.

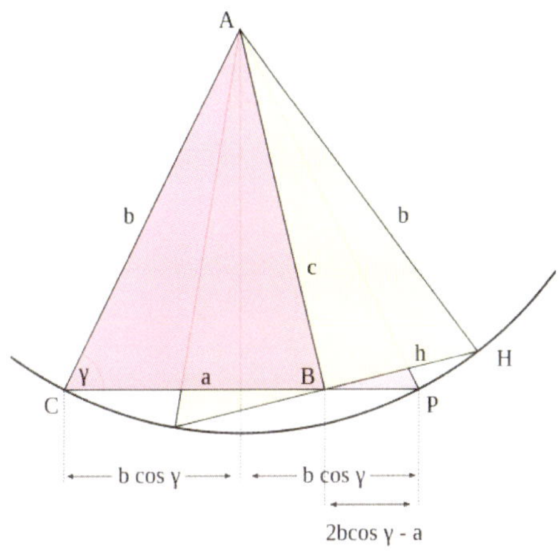

Figure: The triangle ABC (pink), an auxiliary circle (light blue) and two auxiliary right triangles (yellow)

Case of acute angle γ, where $a < 2b\cos\gamma$. Drop the perpendicular from A onto $a = BC$, creating a line segment of length $b\cos\gamma$. Duplicate the right triangle to form the isosceles triangle ACP. Construct the circle with center A and radius b, and a chord through B perpendicular to $c = AB$, half of which is $h = BH$. Apply the Pythagorean theorem to obtain

$$b^2 = c^2 + h^2.$$

Now use the *chord theorem* (Euclid's *Elements*: Book 3, Proposition 35), which says that if two chords intersect, the product of the two line segments obtained on one chord is equal to the product of the two line segments obtained on the other chord. In the present case: $BH^2 = BC \cdot BP$, or

$$h^2 = a(2b\cos\gamma - a).$$

Substituting into the previous equation gives the law of cosines:

$$b^2 = c^2 + a(2b\cos\gamma - a).$$

Note that the power of the point *B* with respect to the circle has the negative value $-h^2$.

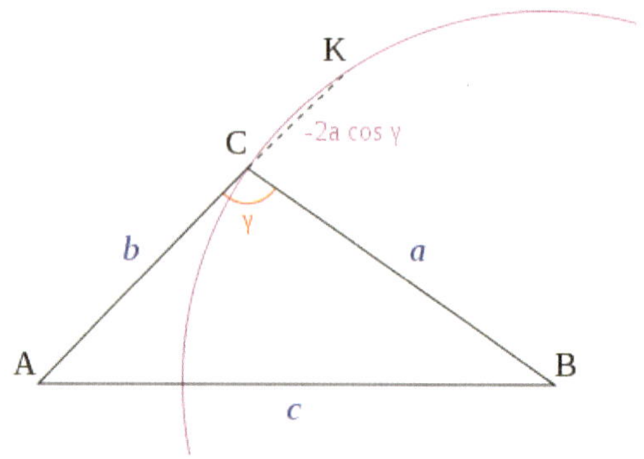

Figure: Proof of the law of cosines using the power of a point theorem.

Case of obtuse angle γ this proof uses the power of a point theorem directly, without the auxiliary triangles obtained by constructing a tangent or a chord. Construct a circle with center *B* and radius *a* (see Figure), which intersects the secant through *A* and *C* in *C* and *K*. The power of the point *A* with respect to the circle is equal to both $AB^2 - BC^2$ and *AC*·*AK*. Therefore,

$$\begin{aligned} c^2 - a^2 &= b(b + 2a\cos(\pi - \gamma)) \\ &= b(b - 2a\cos\gamma), \end{aligned}$$

which is the law of cosines.

Using algebraic measures for line segments (allowing negative numbers as lengths of segments) the case of obtuse angle ($CK > 0$) and acute angle ($CK < 0$) can be treated simultaneously.

Using the Law of Sines

By using the law of sines and knowing that the angles of a triangle must sum to 180 degrees, we have the following system of equations (the three unknowns are the angles):

$$\frac{c}{\sin\gamma} = \frac{b}{\sin\beta},$$

$$\frac{c}{\sin\gamma} = \frac{a}{\sin\alpha},$$

$$\alpha + \beta + \gamma = \pi.$$

Then, by using the third equation of the system, we obtain a system of two equations in two variables:

$$\frac{c}{\sin\gamma} = \frac{b}{\sin(\alpha + \gamma)},$$

$$\frac{c}{\sin\gamma} = \frac{a}{\sin\alpha},$$

where we have used the trigonometric property that the sine of a supplementary angle is equal to the sine of the angle.

Using the identity (see Angle sum and difference identities)

$$\sin(\alpha+\gamma)=\sin\alpha\cos\gamma+\sin\gamma\cos\alpha$$

Leads to

$$c(\sin\alpha\cos\gamma+\sin\gamma\cos\alpha)=b\ \sin\gamma,$$

$$c\sin\alpha=a\sin\gamma.$$

By dividing the whole system by cos γ, we have:

$$c(\sin\alpha+\tan\gamma\cos\alpha)=b\tan\gamma,$$

$$\frac{c\sin\alpha}{\cos\gamma}=a\tan\gamma,$$

$$\frac{c^2\sin^2\alpha}{\cos^2\gamma}=a^2\tan^2\gamma.$$

Hence, from the first equation of the system, we can obtain

$$\frac{c\sin\alpha}{b-c\cos\alpha}=\tan\gamma$$

By substituting this expression into the second equation and by using

$$1+\tan^2\gamma=\frac{1}{\cos^2\gamma}$$

we can obtain one equation with one variable:

$$c^2\sin^2\alpha\left(1+\frac{c^2\sin^2\alpha}{(b-c\cos\alpha)^2}\right)=a^2\cdot\frac{c^2\sin^2\alpha}{(b-c\cos\alpha)^2}$$

By multiplying by $(b - c\cos\alpha)^2$, we can obtain the following equation:

$$(b-c\cos\alpha)^2+c^2\sin^2\alpha=a^2.$$

This implies

$$b^2-2bc\ \cos\alpha+c^2\cos^2\alpha+c^2\sin^2\alpha=a^2.$$

Recalling the Pythagorean identity, we obtain the law of cosines:

$$a^2=b^2+c^2-2bc\cos\alpha.$$

Isosceles Case

When $a = b$, i.e., when the triangle is isosceles with the two sides incident to the angle γ equal, the law of cosines simplifies significantly. Namely, because $a^2 + b^2 = 2a^2 = 2ab$, the law of cosines becomes

$$\cos\gamma = 1 - \frac{c^2}{2a^2}$$

or

$$c^2 = 2a^2(1 - \cos\gamma).$$

Analog for Tetrahedra

An analogous statement begins by taking $\alpha, \beta, \gamma, \delta$ to be the areas of the four faces of a tetrahedron. Denote the dihedral angles by $\widehat{\beta\gamma}$ etc. Then

$$\alpha^2 = \beta^2 + \gamma^2 + \delta^2 - 2\left(\beta\gamma\cos\left(\widehat{\beta\gamma}\right) + \gamma\delta\cos\left(\widehat{\gamma\delta}\right) + \delta\beta\cos\left(\widehat{\delta\beta}\right)\right).$$

Version Suited to Small Angles

When the angle, γ, is small and the adjacent sides, a and b, are of similar length, the right hand side of the standard form of the law of cosines can lose a lot of accuracy to numerical loss of significance. In situations where this is an important concern, a mathematically equivalent version of the law of cosines, similar to the haversine formula, can prove useful:

$$\begin{aligned} c^2 &= (a-b)^2 + 4ab\sin^2\left(\frac{\gamma}{2}\right) \\ &= (a-b)^2 + 4ab\ \text{haversin}(\gamma). \end{aligned}$$

In the limit of an infinitesimal angle, the law of cosines degenerates into the circular arc length formula, $c = a\,\gamma$.

Law of Cosines in non-Euclidean Geometry

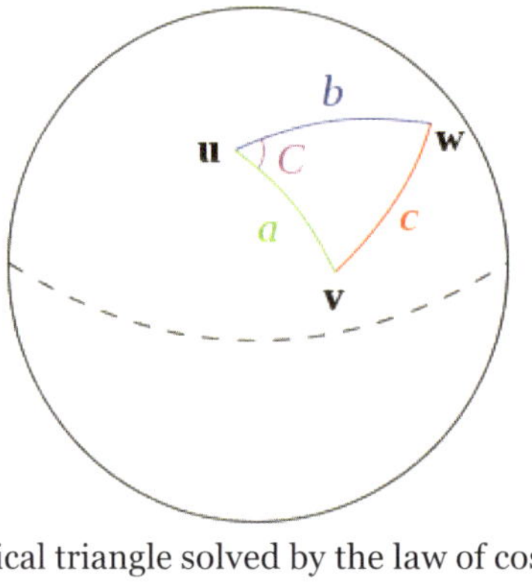

Spherical triangle solved by the law of cosines.

A version of the law of cosines also holds in non-Euclidean geometry. In spherical geometry, a triangle is defined by three points u, v, and w on the unit sphere, and the arcs of great circles connecting those points. If these great circles make angles *A*, *B*, and *C* with opposite sides *a*, *b*, *c* then the spherical law of cosines asserts that both of the following relationships hold:

$$\cos a = \cos b \cos c + \sin b \sin c \cos A$$

$$\cos A = -\cos B \cos C + \sin B \sin C \cos a.$$

In hyperbolic geometry, a pair of equations are collectively known as the hyperbolic law of cosines. The first is

$$\cosh a = \cosh b \, \cosh c - \sinh b \, \sinh c \, \cos A$$

where sinh and cosh are the hyperbolic sine and cosine, and the second is

$$\cos A = -\cos B \cos C + \sin B \sin C \cosh a.$$

As in Euclidean geometry, one can use the law of cosines to determine the angles *A*, *B*, *C* from the knowledge of the sides *a*, *b*, *c*. However, unlike Euclidean geometry, the reverse is also possible in each of the models of non-Euclidean geometry: the angles *A*, *B*, *C* determine the sides *a*, *b*, *c*.

Applications

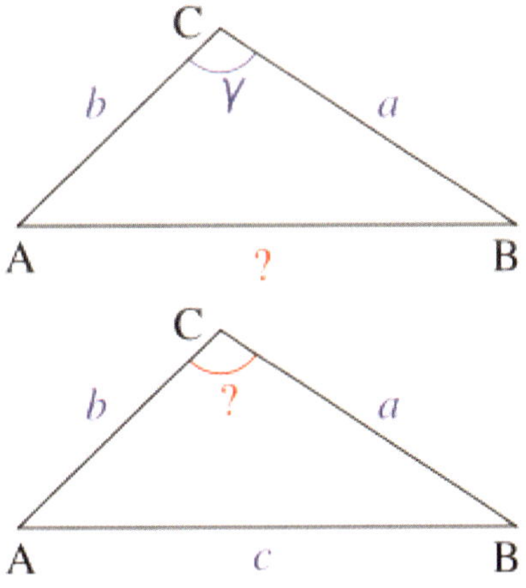

Figure: Applications of the law of cosines: unknown side and unknown angle.

The theorem is used in triangulation, for solving a triangle or circle, i.e., to find:

- the third side of a triangle if one knows two sides and the angle between them:

$$c = \sqrt{a^2 + b^2 - 2ab\cos\gamma};$$

- the angles of a triangle if one knows the three sides:

$$\gamma = \arccos\left(\frac{a^2 + b^2 - c^2}{2ab}\right);$$

- the third side of a triangle if one knows two sides and an angle opposite to one of them (one may also use the Pythagorean theorem to do this if it is a right triangle):

$$a = b\cos\gamma \pm \sqrt{c^2 - b^2\sin^2\gamma}.$$

These formulas produce high round-off errors in floating point calculations if the triangle is very acute, i.e., if c is small relative to a and b or γ is small compared to 1. It is even possible to obtain a result slightly greater than one for the cosine of an angle.

The third formula shown is the result of solving for a in the quadratic equation $a^2 - 2ab\cos\gamma + b^2 - c^2 = 0$. This equation can have 2, 1, or 0 positive solutions corresponding to the number of possible triangles given the data. It will have two positive solutions if $b\sin\gamma < c < b$, only one positive solution if $c = b\sin\gamma$, and no solution if $c < b\sin\gamma$ or $c \geq b$. These different cases are also explained by the side-side-angle congruence ambiguity.

Spherical Law of Cosines

Consider a spherical triangle with sides α, β, and γ, angle Γ opposite γ,. To compute γ, we have the formula

$$cos(\gamma) = cos(\alpha)cos(\beta) + sin(\alpha)sin(\beta)cos(\Gamma)$$

Proof: Project the triangle onto the plane tangent to the sphere a Γ and compute the length of the projection of γ, in two different ways. First, using the plane Law of Cosines in the plane tangent to the sphere at Γ, we see that the length of the projection of γ, is

$$tan^2(\alpha) + tan^2(\beta) - 2tan(\alpha)tan(\beta)cos(\Gamma)$$

Whereas if we use the plane Law of Cosines in the plane containing the great circle of γ, we get that the length of the projection of γ, is

$$sec^2(\alpha) + sec^2(\beta) - 2sec(\alpha)sec(\beta)cos(\gamma)$$

By applying Figure to α and β, In below Figure illustrates these two methods of computing the length of the projection of γ onto the plane tangent at Γ, that is, the red segment:

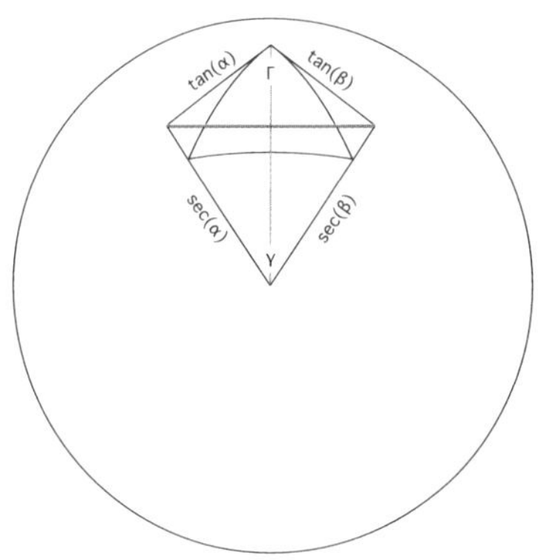

Figure: Two ways to measure the red segment

Subtracting equation $tan^2(\alpha)+tan^2(\beta) - 2tan(\alpha)tan(\beta)cos(\Gamma)$ from equation

$sec^2(\alpha) + sec^2(\beta) - 2sec(\alpha)sec(\beta)cos(\gamma)$, we get that

$$0 = 2 + 2tan(\alpha)tan(\beta)cos(\Gamma) - 2sec(\alpha)sec(\beta)cos(\gamma)$$

Solving for cos(γ), gives The Spherical Law of Cosines:

$$cos(\gamma) = cos(\alpha)cos(\beta) + sin(\alpha)sin(\beta)cos(\Gamma)$$

Corollary: Given the spherical triangle ABΓ with opposing sides α, β, and γ, we have the following:

$$sin(\alpha)cos(B) = cos(\beta)sin(\gamma) - sin(\beta)cos(\gamma)cos(A)$$

Proof: Extend the side γ to $\frac{\pi}{2}$ radians as in below Figure:

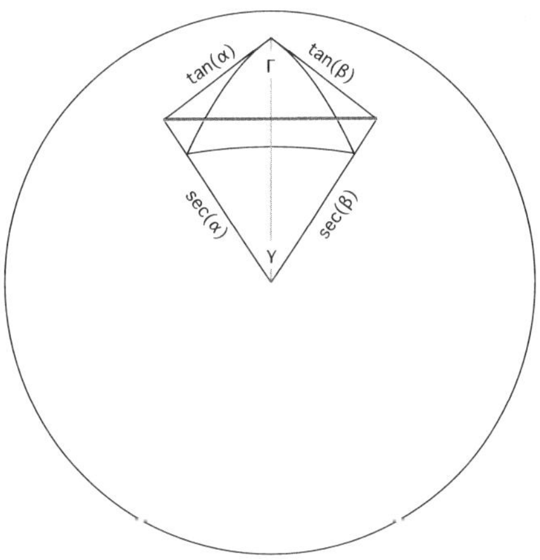

Using The Spherical Law of Cosines, there are two ways of computing cos(δ):

$$\begin{aligned} cos(\delta) &= cos(\alpha)cos(\pi/2) + sin(\alpha)sin(\pi/2)cos(B) \\ &= sin(\alpha)cos(B) \end{aligned}$$

$$\begin{aligned} cos(\delta) &= cos(\beta)cos(\pi/2 - c) + sin(\beta)sin(\pi/2 - c)cos(\pi - A) \\ &= cos(\beta)sin(\gamma) - sin(\beta)cos(\gamma)cos(A) \end{aligned}$$

Equating $\begin{aligned} cos(\delta) &= cos(\alpha)cos(\pi/2) + sin(\alpha)sin(\pi/2)cos(B) \\ &= sin(\alpha)cos(B) \end{aligned}$ and

$$\begin{aligned} cos(\delta) &= cos(\beta)cos(\pi/2 - c) + sin(\beta)sin(\pi/2 - c)cos(\pi - A) \\ &= cos(\beta)sin(\gamma) - sin(\beta)cos(\gamma)cos(A) \end{aligned}$$

we get the corollary:

$$sin(\alpha)cos(B) = cos(\beta)sin(\gamma) - sin(\beta)cos(\gamma)cos(A)$$

Rearrangements

The first and second spherical laws of cosines can be rearranged to put the sides (a, b, c) and angles (A, B, C) on opposite sides of the equations:

$$\cos C = \frac{\cos c - \cos a \cos b}{\sin a \, \sin b}$$

$$\cos c = \frac{\cos C + \cos A \cos B}{\sin A \, \sin B}$$

Planar Limit: Small Angles

For *small* spherical triangles, i.e. for small a, b, and c, the spherical law of cosines is approximately the same as the ordinary planar law of cosines,

$$c^2 \approx a^2 + b^2 - 2ab\cos C.$$

To prove this, we will use the small-angle approximation obtained from the Maclaurin series for the cosine and sine functions:

$$\cos a = 1 - \frac{a^2}{2} + O(a^4), \sin a = a + O(a^3)$$

Substituting these expressions into the spherical law of cosines nets:

$$1 - \frac{c^2}{2} + O(c^4) = 1 - \frac{a^2}{2} - \frac{b^2}{2} + \frac{a^2b^2}{4} + O(a^4) + O(b^4) + \cos(C)(ab + O(a^3b) + O(ab^3) + O(a^3b^3))$$

or after simplifying:

$$c^2 = a^2 + b^2 - 2ab\cos C + O(c^4) + O(a^4) + O(b^4) + O(a^2b^2) + O(a^3b) + O(ab^3) + O(a^3b^3).$$

The big O terms for a and b are dominated by $O(a^4) + O(b^4)$ as a and b get small, so we can write this last expression as:

$$c^2 = a^2 + b^2 - 2ab \, \cos C + O(a^4) + O(b^4) + O(c^4).$$

Spherical Law of Cosines for Angles

We know that Angle-Angle-Angle is a congruence relation for spherical geometry. We can use the Spherical Law of Cosines for Angles to prove this.

We again assume that our triangle XYZ is on a unit sphere, with angles α, β, γ and side lengths a, b, c, with corresponding central angles A, B, C.

Since each line segment on the sphere a part of a great circle, we define the polar point of a line segment to be the center of the great circle it lies on. Since there are two possible choices, we make the choice well defined using the right-hand rule: when moving from X to Y along the line segment, the center should be on the left.

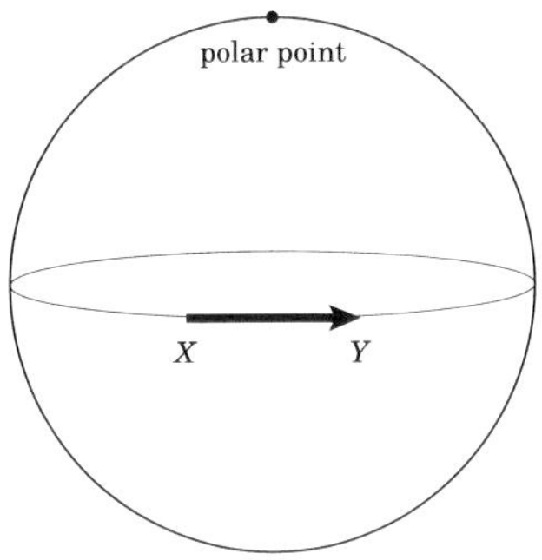

The dual triangle of XYZ is the triangle whose vertices are the polar points of the sides of XYZ. Draw several triangles on the Lénárt sphere and their duals.

Proposition Given a triangle with side lengths A, B, C (measured as central angles) and opposite angles α, β, γ, the dual triangle formed from the polar points has side lengths $\pi-\alpha, \pi-\beta, \pi-\gamma$ (measured as central angles) and angles $\pi-A, \pi-B, \pi-C$. Moreover, the dual of the dual triangle is the original triangle.

Applying the Spherical Law of Cosines to the dual triangle, we obtain

Theorem (Spherical Law of Cosines for Angles). $\cos\gamma = (\cos\alpha)(\cos\beta) + (\sin\alpha)(\sin\beta)\cos C$.

Law of Tangents

Laws of tangent represent the relation between difference and sum of sides of a right triangle and tangents of half of difference and sum of corresponding angles. The formulas for laws of tangent are illustrated below.

Let us assume a right triangle ABC in which side opposite to ∠A, ∠B and ∠C are a, b and c respectively. Then, according to laws of tangent, we have following three relations:

$$\frac{a-b}{a+b}=\frac{\tan\frac{A-B}{2}}{\tan\frac{A+B}{2}}$$

Similarly, for other sides:

$$\frac{b-c}{b+c}=\frac{\tan\frac{B-C}{2}}{\tan\frac{B+C}{2}}$$

And

$$\frac{c-a}{c+a}=\frac{\tan\frac{C-A}{2}}{\tan\frac{C+A}{2}}$$

Proof

Let us have a look at the proof of laws of tangent.

We have law of sine :

$$\frac{a}{\sin A} = \frac{b}{\sin B} = \frac{c}{\sin C}$$

Lets use first and second relation

$$\frac{a}{sin\ A} = \frac{b}{sin\ B} = k(say)$$

$$a = k \sin A \ and\ b = k \sin B$$
$$a - b = k(sin\ A - sin\ B)$$
$$a + b = k(sin\ A + sin\ B)$$

So,

$$\frac{a-b}{a+b} = \frac{sin\ A - sin\ B}{sin\ A + sin\ B}$$

We have the following identities:

$$sin\ A + sin\ B = \sin\left(\frac{A+B}{2}\right)\cos\left(\frac{A-B}{2}\right)$$

$$sin\ A - sin\ B = \sin\left(\frac{A-B}{2}\right)\sin\left(\frac{A+B}{2}\right)$$

Substituting those in equation $\frac{a-b}{a+b} = \frac{sin\ A - sin\ B}{sin\ A + sin\ B}$:

$$\frac{a-b}{a+b} = \frac{\sin\left(\frac{A-B}{2}\right)\cos\left(\frac{A+B}{2}\right)}{\sin\left(\frac{A+B}{2}\right)\cos\left(\frac{A-B}{2}\right)}$$

$$\frac{a-b}{a+b} = \frac{\tan\left(\frac{A-B}{2}\right)}{\tan\left(\frac{A+B}{2}\right)}$$

Which is first law of tangent In the similar way (using other relations from laws of sine), other two laws of tangent can be proved.

Solve Problems

There are many different types of problems which may be solved using laws of tangent. Let us go ahead and understand how to solve problems with laws of tangent.

These laws can be used in order to find the missing parts of the triangle.

For Example: One side and two angles are given. We need to estimate the length of third side of right triangles. In this case, we may use law of tangent as shown below:

$$\frac{a-b}{a+b}=\frac{\tan\left(\frac{A-B}{2}\right)}{\tan\left(\frac{A+B}{2}\right)}$$

$$a=b+(a+b)\left[\frac{\tan\left(\frac{A-B}{2}\right)}{\tan\left(\frac{A+B}{2}\right)}\right]$$

When to find sum of two sides or difference of two sides and corresponding angles are given, laws of tangent can be utilized. Similarly, in order to find sum or difference of two angles when two corresponding sides are given, one can use laws of tangent.

Also, when tangent of sum or difference of two angles is given and sum or differences of lengths of two sides are to be estimated; we may use these laws of tangent.

Problems

Few problems based on laws of tangent are given below:

Problem 1: If in a triangle ABC, $\angle B = 90^{\circ}$, $\angle C = 30^{\circ}$. If side opposite to $\angle B$ is 5 cm. Find the value of side opposite to $\angle C$.

Solution: $$\frac{b-c}{b+c}=\frac{\tan\frac{B-C}{2}}{\tan\frac{B+C}{2}}$$

$$\frac{5-c}{5+c}=\frac{\tan\frac{90-30}{2}}{\tan\frac{90+30}{2}}$$

$$\frac{5-c}{5+c}=\frac{\tan 30}{\tan 60}$$

$$\frac{5-c}{5+c}=\frac{\frac{1}{\sqrt{3}}}{\sqrt{3}}$$

$$\frac{5-c}{5+c}=\frac{1}{3}$$

$$(5-c)3=5+c$$

$$15-3c=5+c$$

$$4c=10$$

$$c=2.5$$

The required length is 2.5 cm.

Problem: Let us consider a triangle ABC in which $\angle C = 60^\circ$. The lengths of sides opposite to $\angle A$ and $\angle B$ are 3 cm and 2 cm respectively. Find the measures of other two angles.

Solution: We know that

$$\angle A+\angle B+\angle C=180^\circ$$

$$\angle A+\angle B=180^\circ-\angle C$$

$$\angle A+\angle B=120^\circ$$

$$\frac{a-b}{a+b}=\frac{\tan\left(\frac{A-B}{2}\right)}{\tan\left(\frac{A+B}{2}\right)}$$

$$\frac{3-2}{3+2}=\frac{\tan\left(\frac{A-B}{2}\right)}{\tan\left(\frac{120^0}{2}\right)}$$

$$\frac{1}{5}=\frac{\tan\left(\frac{A-B}{2}\right)}{\tan 60^0}$$

$$\frac{1}{5}=\frac{\tan\left(\frac{A-B}{2}\right)}{\sqrt{3}}$$

$$0.3464=\tan\left(\frac{A-B}{2}\right)$$

$$\frac{A-B}{2}=\tan^{-1}0.3464$$

$$\frac{A-B}{2}=19.11$$

$$\angle A-\angle B=38.22^\circ$$

$\angle DAB = \angle CBD$

$\angle DAB = \angle BCD$

$\angle CDB = \angle ADB = 90^{0}$

Using the AA criterion for the similarity of triangles, $\Delta BDC \cong \Delta ABC$

Thus, it can be concluded $\Delta ADB \cong \Delta BDC$

So if a perpendicular is drawn from the right-angled vertex of a right triangle to the hypotenuse, then the triangles formed on both sides of the perpendicular are similar to each other and also to the whole triangle.

Now, we are required to prove $AC^2 = AB^2 + BC^2$. We drop a perpendicular BD on the side AC.

We already know that $\Delta ADB \cong \Delta ABC$

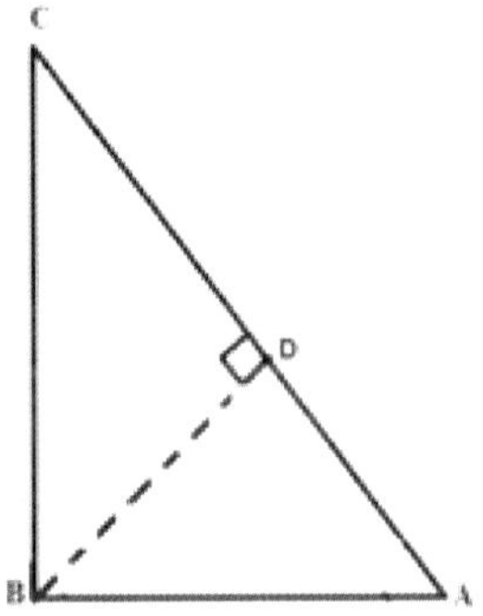

$\therefore \dfrac{AD}{AB} = \dfrac{AB}{BC}$ (Condition for similarity) Or $AB^2 = AC \times AD$

Also, $\Delta BDC \cong \Delta ABC$

$\therefore \dfrac{CD}{BC} = \dfrac{BC}{AC}$ (Condition for similarity) or $BC^2 = AC \times CD$

Summing above in equations,

$AB^2 + BC^2 = CD.AC + AD.AC$

$\Rightarrow AB^2 + BC^2 = AC(CD + AD)$

However, $CD + AD = AC$

Thus, $AC^2 = AB^2 + BC^2$

Hence, the proof of the Pythagoras theorem.

Other Forms of the Theorem

As pointed out in the introduction, if c denotes the length of the hypotenuse and a and b denote the lengths of the other two sides, the Pythagorean theorem can be expressed as the Pythagorean equation:

$$a^2 + b^2 = c^2.$$

If the length of both a and b are known, then c can be calculated as

$$c = \sqrt{a^2 + b^2}.$$

If the length of the hypotenuse c and of one side (a or b) are known, then the length of the other side can be calculated as

$$a = \sqrt{c^2 - b^2}$$

or

$$b = \sqrt{c^2 - a^2}.$$

The Pythagorean equation relates the sides of a right triangle in a simple way, so that if the lengths of any two sides are known the length of the third side can be found. Another corollary of the theorem is that in any right triangle, the hypotenuse is greater than any one of the other sides, but less than their sum.

A generalization of this theorem is the law of cosines, which allows the computation of the length of any side of any triangle, given the lengths of the other two sides and the angle between them. If the angle between the other sides is a right angle, the law of cosines reduces to the Pythagorean equation

Euclid's Proof

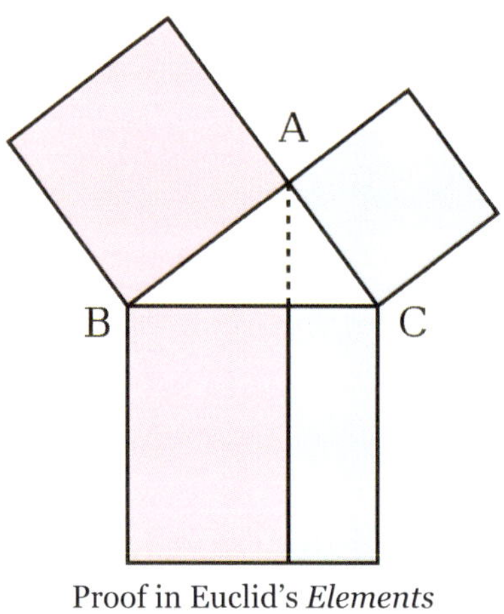

Proof in Euclid's *Elements*

In outline, here is how the proof in Euclid's *Elements* proceeds. The large square is divided into a left and right rectangle. A triangle is constructed that has half the area of the left rectangle. Then another triangle is constructed that has half the area of the square on the left-most side. These two triangles are shown to be congruent, proving this square has the same area as the left rectangle. This argument is followed by a similar version for the right rectangle and the remaining square.

Putting the two rectangles together to reform the square on the hypotenuse, its area is the same as the sum of the area of the other two squares. The details follow.

Let *A*, *B*, *C* be the vertices of a right triangle, with a right angle at *A*. Drop a perpendicular from *A* to the side opposite the hypotenuse in the square on the hypotenuse. That line divides the square on the hypotenuse into two rectangles, each having the same area as one of the two squares on the legs.

For the formal proof, we require four elementary lemmata:

1. If two triangles have two sides of the one equal to two sides of the other, each to each, and the angles included by those sides equal, then the triangles are congruent (side-angle-side).
2. The area of a triangle is half the area of any parallelogram on the samc base and having the same altitude.
3. The area of a rectangle is equal to the product of two adjacent sides.
4. The area of a square is equal to the product of two of its sides (follows from 3).

Next, each top square is related to a triangle congruent with another triangle related in turn to one of two rectangles making up the lower square.

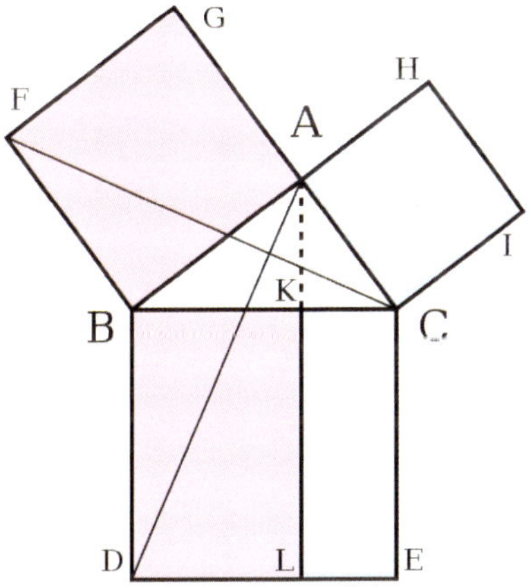

Illustration including the new lines

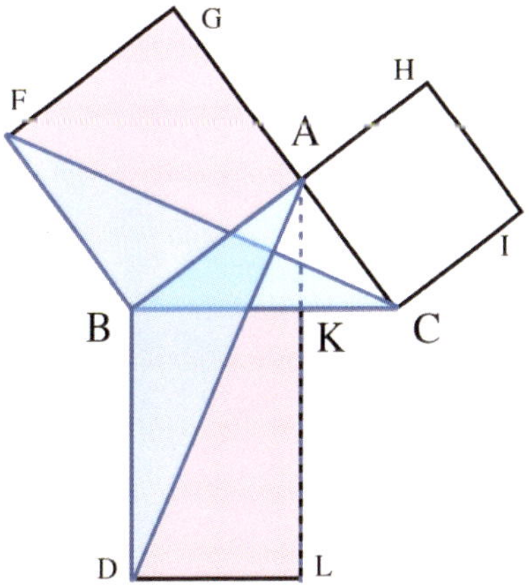

Showing the two congruent triangles of half the area of rectangle BDLK and square BAGF

The proof is as follows:

1. Let ACB be a right-angled triangle with right angle CAB.
2. On each of the sides BC, AB, and CA, squares are drawn, CBDE, BAGF, and ACIH, in that order. The construction of squares requires the immediately preceding theorems in Euclid, and depends upon the parallel postulate.
3. From A, draw a line parallel to BD and CE. It will perpendicularly intersect BC and DE at K and L, respectively.
4. Join CF and AD, to form the triangles BCF and BDA.
5. Angles CAB and BAG are both right angles; therefore C, A, and G are collinear. Similarly for B, A, and H.
6. Angles CBD and FBA are both right angles; therefore angle ABD equals angle FBC, since both are the sum of a right angle and angle ABC.
7. Since AB is equal to FB and BD is equal to BC, triangle ABD must be congruent to triangle FBC.
8. Since A-K-L is a straight line, parallel to BD, then rectangle BDLK has twice the area of triangle ABD because they share the base BD and have the same altitude BK, i.e., a line normal to their common base, connecting the parallel lines BD and AL. (lemma 2)
9. Since C is collinear with A and G, square BAGF must be twice in area to triangle FBC.
10. Therefore, rectangle BDLK must have the same area as square BAGF = AB^2.
11. Similarly, it can be shown that rectangle CKLE must have the same area as square ACIH = AC^2.
12. Adding these two results, $AB^2 + AC^2 = BD \times BK + KL \times KC$
13. Since BD = KL, $BD \times BK + KL \times KC = BD(BK + KC) = BD \times BC$
14. Therefore, $AB^2 + AC^2 = BC^2$, since CBDE is a square.

This proof, which appears in Euclid's *Elements* as that of Proposition 47 in Book 1, demonstrates that the area of the square on the hypotenuse is the sum of the areas of the other two squares. This is quite distinct from the proof by similarity of triangles, which is conjectured to be the proof that Pythagoras used.

Proofs by Dissection and Rearrangement

We have already discussed the Pythagorean proof, which was a proof by rearrangement. The same idea is conveyed by the leftmost figure, which consists of a large square, side $a + b$, containing four identical right triangles. The triangles are shown in two arrangements, the first of which leaves two squares a^2 and b^2 uncovered, the second of whom leaves square c^2 uncovered. The area encompassed by the outer square never changes, and the area of the four triangles is the same at the beginning and the end, so the black square areas must be equal, therefore $a^2 + b^2 = c^2$.

A second proof by rearrangement is given by the middle figure. A large square is formed with area c^2, from four identical right triangles with sides a, b and c, fitted around a small central square. Then two rectangles are formed with sides a and b by moving the triangles. Combining the smaller square with these rectangles produces two squares of areas a^2 and b^2, which must have the same area as the initial large square.

The third, rightmost image also gives a proof. The upper two squares are divided as shown by the blue and green shading, into pieces that when rearranged can be made to fit in the lower square on the hypotenuse – or conversely the large square can be divided as shown into pieces that fill the other two. This way of cutting one figure into pieces and rearranging them to get another figure is called dissection. This shows the area of the large square equals that of the two smaller ones.

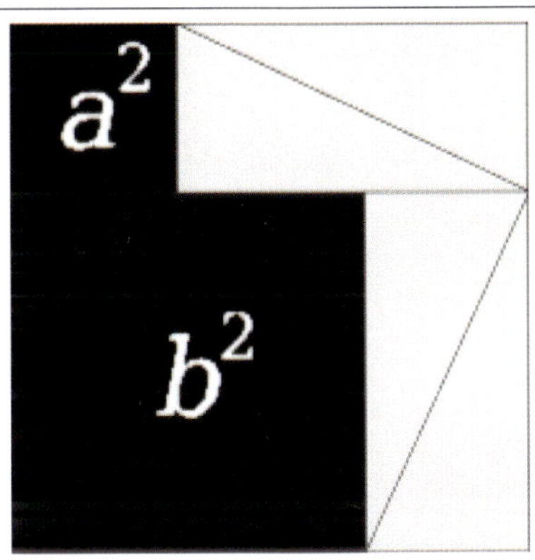	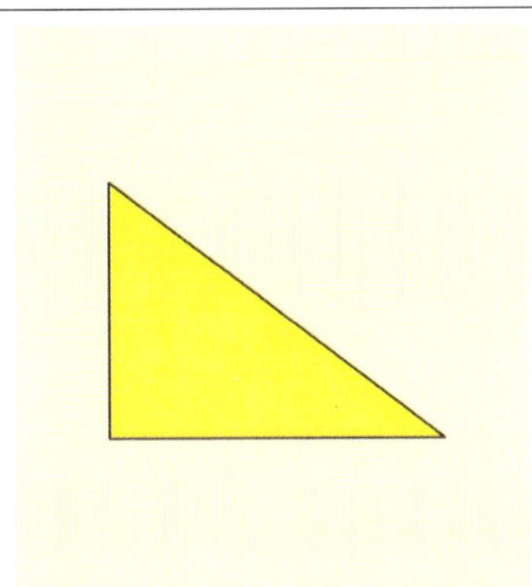	
Figure showing proof by rearrangement of four identical right triangles	Figure showing another proof by rearrangement	Proof using an elaborate rearrangement

Einstein's Proof by Dissection without Rearrangement

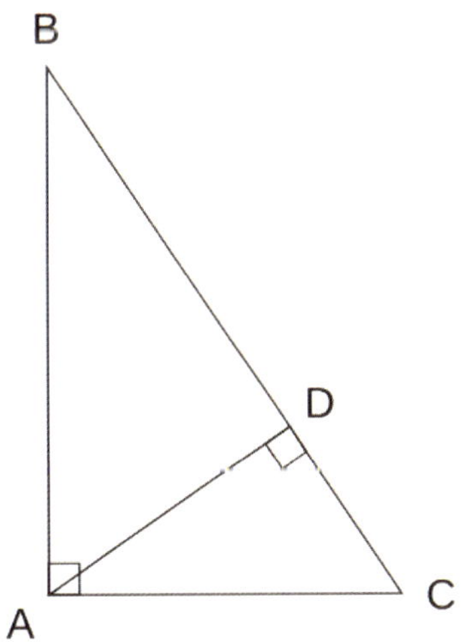

Right triangle on the hypotenuse dissected into two similar right triangles on the legs, according to Einstein's proof

Albert Einstein gave a proof by dissection in which the pieces need not get moved. Instead of using a square on the hypotenuse and two squares on the legs, one can use any other shape that includes the hypotenuse, and two similar shapes that each include one of two legs instead of the hypotenuse. In Einstein's proof, the shape that includes the hypotenuse is the right triangle itself. The dissection consists of dropping a perpendicular from the vertex of the right angle of the triangle to the hypotenuse, thus splitting the whole triangle into two parts. Those two parts have the same shape as the original right triangle, and have the legs of the original triangle as their hypotenuses, and the sum of their areas is that of the original triangle. Because the ratio of the area of a right triangle to the square of its hypotenuse is the same for similar triangles, the relationship between the areas of the three triangles holds for the squares of the sides of the large triangle as well.

Algebraic Proofs

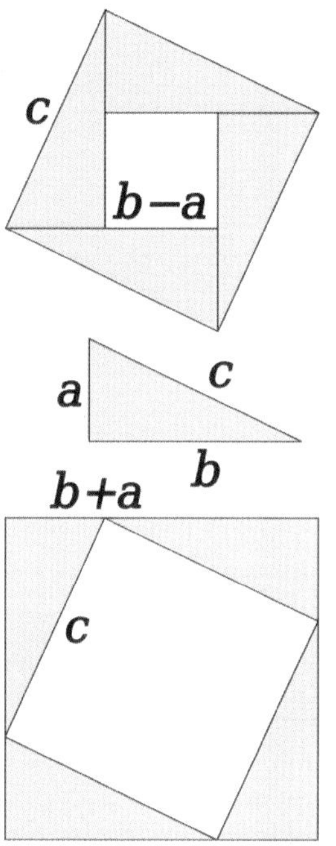

Diagram of the two algebraic proofs

The theorem can be proved algebraically using four copies of a right triangle with sides a, b and c, arranged inside a square with side c as in the top half of the diagram. The triangles are similar with area $\frac{1}{2}ab$ while the small square has side $b - a$ and area $(b - a)^2$. The area of the large square is therefore

$$(b-a)^2 + 4\frac{ab}{2} = (b-a)^2 + 2ab = b^2 - 2ab + a^2 + 2ab = a^2 + b^2.$$

But this is a square with side c and area c^2, so

$$c^2 = a^2 + b^2.$$

A similar proof uses four copies of the same triangle arranged symmetrically around a square with side c, as shown in the lower part of the diagram. This results in a larger square, with side $a + b$ and area $(a + b)^2$. The four triangles and the square side c must have the same area as the larger square,

$$(b+a)^2 = c^2 + 4\frac{ab}{2} = c^2 + 2ab,$$

Giving

$$c^2 = (b+a)^2 - 2ab = b^2 + 2ab + a^2 - 2ab = a^2 + b^2.$$

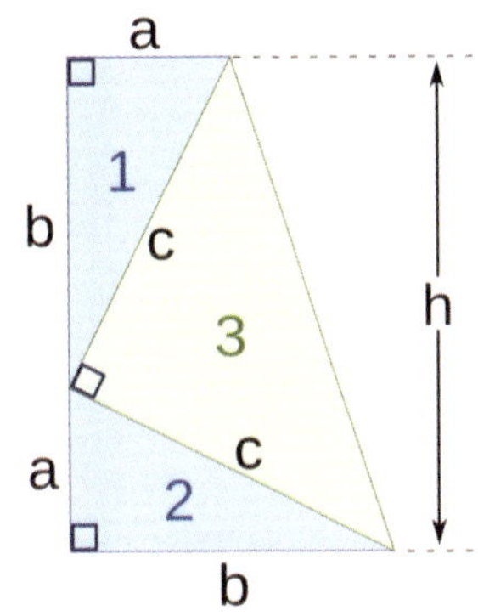

Diagram of Garfield's proof

A related proof was published by future U.S. President James A. Garfield (then a U.S. Representative). Instead of a square it uses a trapezoid, which can be constructed from the square in the second of the above proofs by bisecting along a diagonal of the inner square, to give the trapezoid as shown in the diagram. The area of the trapezoid can be calculated to be half the area of the square, that is

$$\frac{1}{2}(b+a)^2.$$

The inner square is similarly halved, and there are only two triangles so the proof proceeds as above except for a factor of $\frac{1}{2}$, which is removed by multiplying by two to give the result.

Proof using Differentials

One can arrive at the Pythagorean theorem by studying how changes in a side produce a change in the hypotenuse and employing calculus.

The triangle ABC is a right triangle, as shown in the upper part of the diagram, with BC the hypotenuse. At the same time the triangle lengths are measured as shown, with the hypotenuse of length y, the side AC of length x and the side AB of length a, as seen in the lower diagram part.

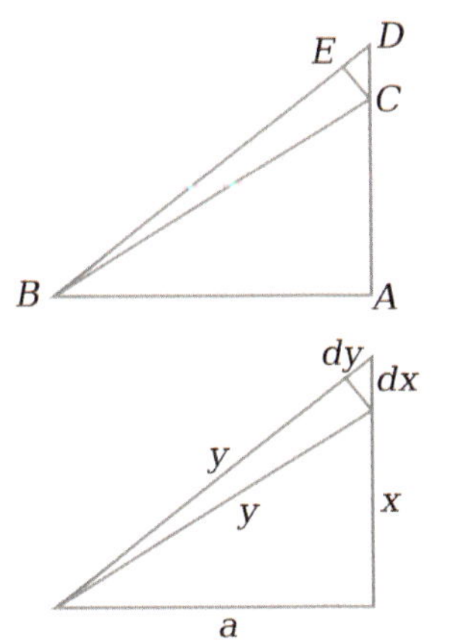

Diagram for differential proof

If x is increased by a small amount dx by extending the side AC slightly to D, then y also increases by dy. These form two sides of a triangle, CDE, which (with E chosen so CE is perpendicular to the hypotenuse) is a right triangle approximately similar to ABC. Therefore, the ratios of their sides must be the same, that is:

$$\frac{dy}{dx} = \frac{x}{y}.$$

This can be rewritten as $ydy = xdx$, which is a differential equation that can be solved by direct integration:

$$\int ydy = \int xdx,$$

Giving

$$y^2 = x^2 + C.$$

The constant can be deduced from $x = 0$, $y = a$ to give the equation

$$y^2 = x^2 + a^2.$$

This is more of an intuitive proof than a formal one: it can be made more rigorous if proper limits are used in place of dx and dy.

Converse

The converse of the theorem is also true:

For any three positive numbers a, b, and c such that $a^2 + b^2 = c^2$, there exists a triangle with sides a, b and c, and every such triangle has a right angle between the sides of lengths a and b.

An alternative statement is:

For any triangle with sides a, b, c, if $a^2 + b^2 = c^2$, then the angle between a and b measures 90°.

This converse also appears in Euclid's *Elements*:

"If in a triangle the square on one of the sides equals the sum of the squares on the remaining two sides of the triangle, then the angle contained by the remaining two sides of the triangle is right."

It can be proven using the law of cosines or as follows:

Let ABC be a triangle with side lengths a, b, and c, with $a^2 + b^2 = c^2$. Construct a second triangle with sides of length a and b containing a right angle. By the Pythagorean theorem, it follows that the hypotenuse of this triangle has length $c = \sqrt{a^2 + b^2}$, the same as the hypotenuse of the first triangle. Since both triangles' sides are the same lengths a, b and c, the triangles are congruent and must have the same angles. Therefore, the angle between the side of lengths a and b in the original triangle is a right angle.

The above proof of the converse makes use of the Pythagorean theorem itself. The converse can also be proven without assuming the Pythagorean theorem.

A corollary of the Pythagorean theorem's converse is a simple means of determining whether a triangle is right, obtuse, or acute, as follows. Let c be chosen to be the longest of the three sides and $a + b > c$ (otherwise there is no triangle according to the triangle inequality). The following statements apply:

- If $a^2 + b^2 = c^2$, then the triangle is right.
- If $a^2 + b^2 > c^2$, then the triangle is acute.
- If $a^2 + b^2 < c^2$, then the triangle is obtuse.

Edsger W. Dijkstra has stated this proposition about acute, right, and obtuse triangles in this language:

$$\operatorname{sgn}(\alpha + \beta - \gamma) = \operatorname{sgn}(a^2 + b^2 - c^2),$$

where α is the angle opposite to side a, β is the angle opposite to side b, γ is the angle opposite to side c, and sgn is the sign function.

Consequences and uses of the Theorem

Pythagorean Triples

A Pythagorean triple has three positive integers a, b, and c, such that $a^2 + b^2 = c^2$. In other words, a Pythagorean triple represents the lengths of the sides of a right triangle where all three sides have integer lengths. Such a triple is commonly written (a, b, c).

A primitive Pythagorean triple is one in which a, b and c are coprime (the greatest common divisor of a, b and c is 1).

The following is a list of primitive Pythagorean triples with values less than 100:

(3, 4, 5), (5, 12, 13), (7, 24, 25), (8, 15, 17), (9, 40, 41), (11, 60, 61), (12, 35, 37), (13, 84, 85), (16, 63, 65), (20, 21, 29), (28, 45, 53), (33, 56, 65), (36, 77, 85), (39, 80, 89), (48, 55, 73), (65, 72, 97)

Incommensurable Lengths

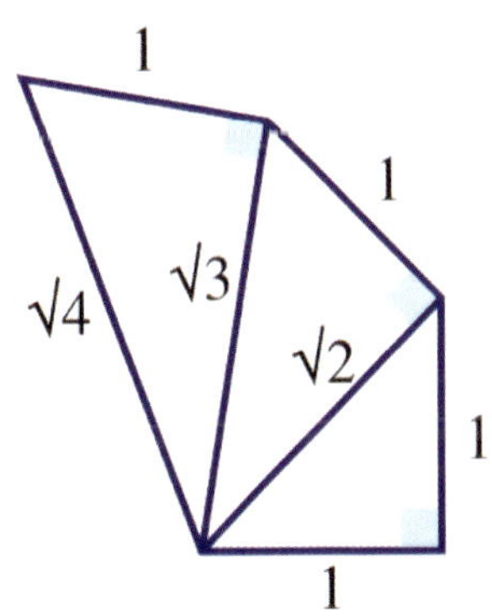

The spiral of Theodorus: A construction for line segments with lengths whose ratios are the square root of a positive integer

One of the consequences of the Pythagorean theorem is that line segments whose lengths are incommensurable (so the ratio of which is not a rational number) can be constructed using a straightedge and compass. Pythagoras's theorem enables construction of incommensurable lengths because the hypotenuse of a triangle is related to the sides by the square root operation.

The figure on the right shows how to construct line segments whose lengths are in the ratio of the square root of any positive integer. Each triangle has a side (labeled "1") that is the chosen unit for measurement. In each right triangle, Pythagoras's theorem establishes the length of the hypotenuse in terms of this unit. If a hypotenuse is related to the unit by the square root of a positive integer that is not a perfect square, it is a realization of a length incommensurable with the unit, such as $\sqrt{2}$, $\sqrt{3}$, $\sqrt{5}$.

Incommensurable lengths conflicted with the Pythagorean school's concept of numbers as only whole numbers. The Pythagorean school dealt with proportions by comparison of integer multiples of a common subunit. According to one legend, Hippasus of Metapontum (*ca.* 470 B.C.) was drowned at sea for making known the existence of the irrational or incommensurable.

Complex Numbers

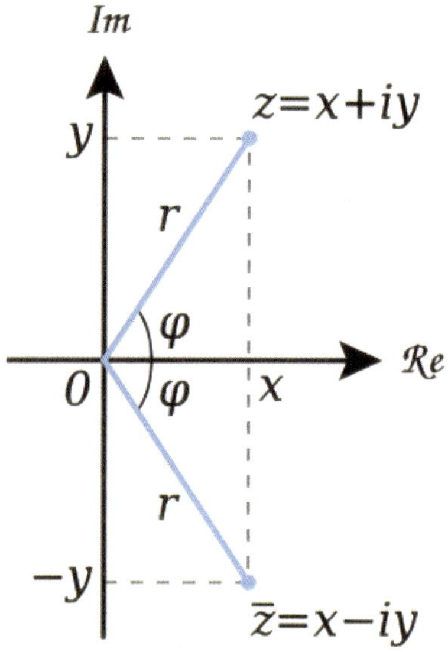

The absolute value of a complex number z is the distance r from z to the origin

For any complex number

$$z = x + iy,$$

the absolute value or modulus is given by

$$r = |z| = \sqrt{x^2 + y^2}.$$

So the three quantities, r, x and y are related by the Pythagorean equation,

$$r^2 = x^2 + y^2.$$

Note that r is defined to be a positive number or zero but x and y can be negative as well as positive. Geometrically r is the distance of the z from zero or the origin O in the complex plane.

This can be generalised to find the distance between two points, z_1 and z_2 say. The required distance is given by

$$|z_1 - z_2| = \sqrt{(x_1 - x_2)^2 + (y_1 - y_2)^2},$$

so again they are related by a version of the Pythagorean equation,

$$|z_1 - z_2|^2 = (x_1 - x_2)^2 + (y_1 - y_2)^2.$$

Euclidean Distance in Various Coordinate Systems

The distance formula in Cartesian coordinates is derived from the Pythagorean theorem. If (x_1, y_1) and (x_2, y_2) are points in the plane, then the distance between them, also called the Euclidean distance, is given by

$$\sqrt{(x_1 - x_2)^2 + (y_1 - y_2)^2}.$$

More generally, in Euclidean n-space, the Euclidean distance between two points, $A=(a_1, a_2, \ldots, a_n)$ and $B=(b_1, b_2, \ldots, b_n)$,, is defined, by generalization of the Pythagorean theorem, as:

$$\sqrt{(a_1 - b_1)^2 + (a_2 - b_2)^2 + \cdots + (a_n - b_n)^2} = \sqrt{\sum_{i=1}^{n} (a_i - b_i)^2}.$$

If Cartesian coordinates are not used, for example, if polar coordinates are used in two dimensions or, in more general terms, if curvilinear coordinates are used, the formulas expressing the Euclidean distance are more complicated than the Pythagorean theorem, but can be derived from it. A typical example where the straight-line distance between two points is converted to curvilinear coordinates can be found in the applications of Legendre polynomials in physics. The formulas can be discovered by using Pythagoras's theorem with the equations relating the curvilinear coordinates to Cartesian coordinates. For example, the polar coordinates (r, θ) can be introduced as:

$$x = r\cos\theta, y = r\sin\theta.$$

Then two points with locations (r_1, θ_1) and (r_2, θ_2) are separated by a distance s:

$$s^2 = (x_1 - x_2)^2 + (y_1 - y_2)^2 = (r_1\cos\theta_1 - r_2\cos\theta_2)^2 + (r_1\sin\theta_1 - r_2\sin\theta_2)^2.$$

Performing the squares and combining terms, the Pythagorean formula for distance in Cartesian coordinates produces the separation in polar coordinates as:

$$\begin{aligned} s^2 &= r_1^2 + r_2^2 - 2r_1r_2\left(\cos\theta_1\cos\theta_2 + \sin\theta_1\sin\theta_2\right) \\ &= r_1^2 + r_2^2 - 2r_1r_2\cos\left(\theta_1 - \theta_2\right) \\ &= r_1^2 + r_2^2 - 2r_1r_2\cos\Delta\theta, \end{aligned}$$

using the trigonometric product-to-sum formulas. This formula is the law of cosines, sometimes called the generalized Pythagorean theorem. From this result, for the case where the radii to the two locations are at right angles, the enclosed angle $\Delta\theta = \pi/2$, and the form corresponding to Pythagoras's theorem is regained: $s^2 = r_1^2 + r_2^2$. The Pythagorean theorem, valid for right triangles, therefore is a special case of the more general law of cosines, valid for arbitrary triangles.

Pythagorean Trigonometric Identity

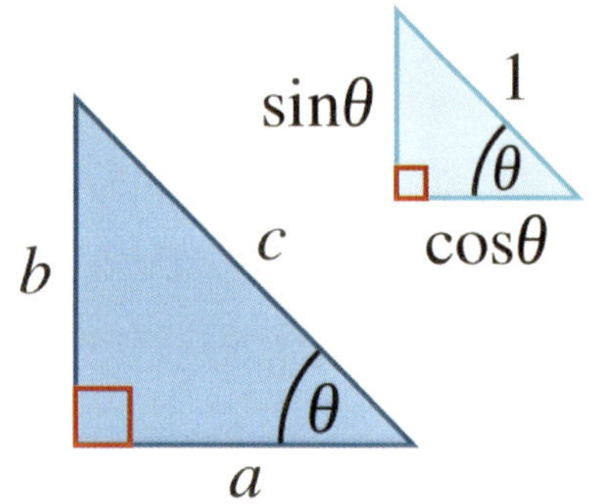

Similar right triangles showing sine and cosine of angle θ

In a right triangle with sides a, b and hypotenuse c, trigonometry determines the sine and cosine of the angle θ between side a and the hypotenuse as:

$$\sin\theta = \frac{b}{c}, \quad \cos\theta = \frac{a}{c}.$$

From that it follows:

$$\cos^2\theta + \sin^2\theta = \frac{a^2 + b^2}{c^2} = 1,$$

where the last step applies Pythagoras's theorem. This relation between sine and cosine is sometimes called the fundamental Pythagorean trigonometric identity. In similar triangles, the ratios of the sides are the same regardless of the size of the triangles, and depend upon the angles. Consequently, in the figure, the triangle with hypotenuse of unit size has opposite side of size $\sin\theta$ and adjacent side of size $\cos\theta$ in units of the hypotenuse.

Relation to the Cross Product

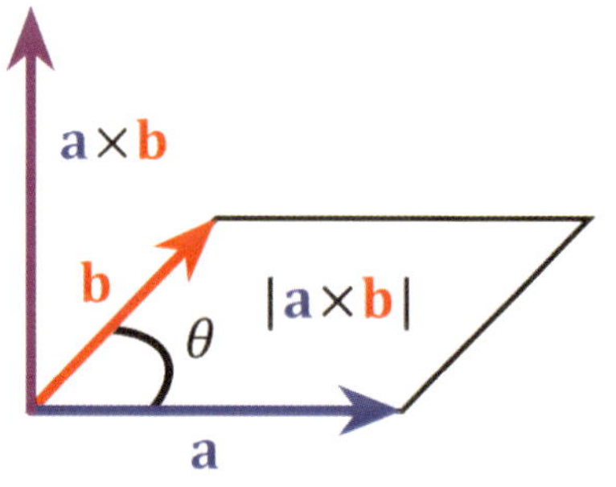

The area of a parallelogram as a cross product; vectors a and b identify a plane and a × b is normal to this plane.

The Pythagorean theorem relates the cross product and dot product in a similar way:

$$\|\mathbf{a}\times\mathbf{b}\|^2+(\mathbf{a}\cdot\mathbf{b})^2=\|\mathbf{a}\|^2\|\mathbf{b}\|^2.$$

This can be seen from the definitions of the cross product and dot product, as

$$\mathrm{a}\times\mathrm{b}=ab\mathrm{n}\sin\theta$$
$$\mathrm{a}\cdot\mathrm{b}=ab\cos\theta,$$

with n a unit vector normal to both a and b. The relationship follows from these definitions and the Pythagorean trigonometric identity.

This can also be used to define the cross product. By rearranging the following equation is obtained

$$\|\mathrm{a}\times\mathrm{b}\|^2=\|\mathrm{a}\|^2\|\mathrm{b}\|^2-(\mathrm{a}\cdot\mathrm{b})^2.$$

This can be considered as a condition on the cross product and so part of its definition, for example in seven dimensions.

Generalizations

Similar figures on the three sides

A generalization of the Pythagorean theorem extending beyond the areas of squares on the three sides to similar figures was known by Hippocrates of Chios in the 5th century BC, and was included by Euclid in his *Elements*:

If one erects similar figures with corresponding sides on the sides of a right triangle, then the sum of the areas of the ones on the two smaller sides equals the area of the one on the larger side.

This extension assumes that the sides of the original triangle are the corresponding sides of the three congruent figures (so the common ratios of sides between the similar figures are $a{:}b{:}c$). While Euclid's proof only applied to convex polygons, the theorem also applics to concave polygons and even to similar figures that have curved boundaries (but still with part of a figure's boundary being the side of the original triangle).

The basic idea behind this generalization is that the area of a plane figure is proportional to the square of any linear dimension, and in particular is proportional to the square of the length of any side. Thus, if similar figures with areas A, B and C are erected on sides with corresponding lengths a, b and c then:

$$\frac{A}{a^2}=\frac{B}{b^2}=\frac{C}{c^2},$$

$$\Rightarrow A+B=\frac{a^2}{c^2}C+\frac{b^2}{c^2}C.$$

But, by the Pythagorean Theorem, $a^2+b^2=c^2$, so $A+B=C$.

Conversely, if we can prove that $A + B = C$ for three similar figures without using the Pythagorean Theorem, then we can work backwards to construct a proof of the theorem. For example, the starting center triangle can be replicated and used as a triangle C on its hypotenuse, and two similar right triangles (A and B) constructed on the other two sides, formed by dividing the central triangle by its altitude. The sum of the areas of the two smaller triangles therefore is that of the third, thus $A + B = C$ and reversing the above logic leads to the Pythagorean Theorem $a^2 + b^2 = c^2$.

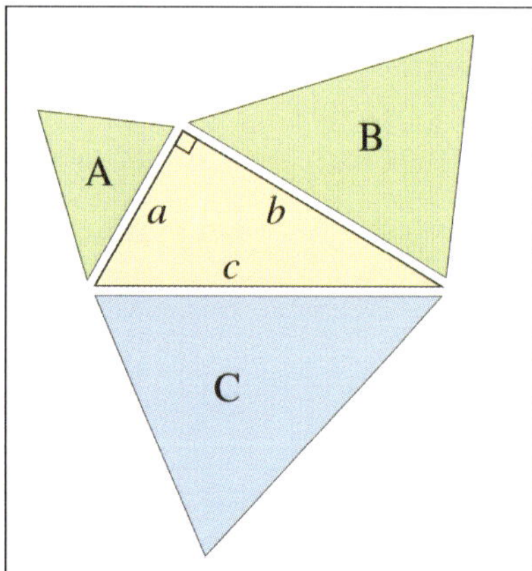

Generalization for similar triangles, green area A + B = blue area C

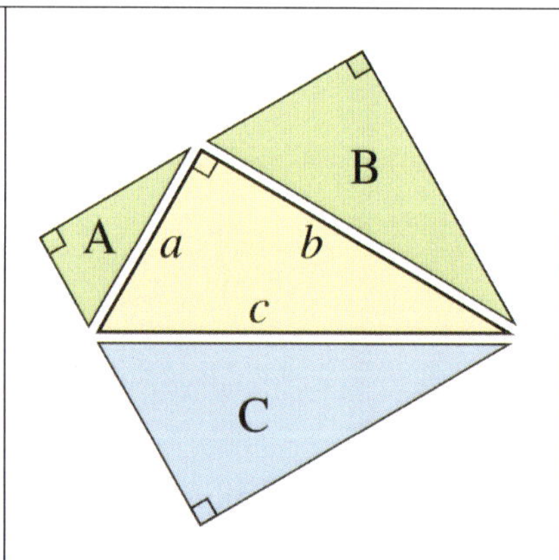

Pythagoras's theorem using similar right triangles

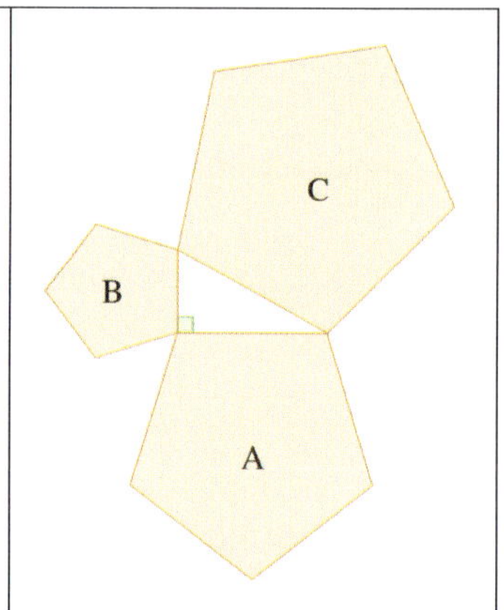

Generalization for regular pentagons

Law of Cosines

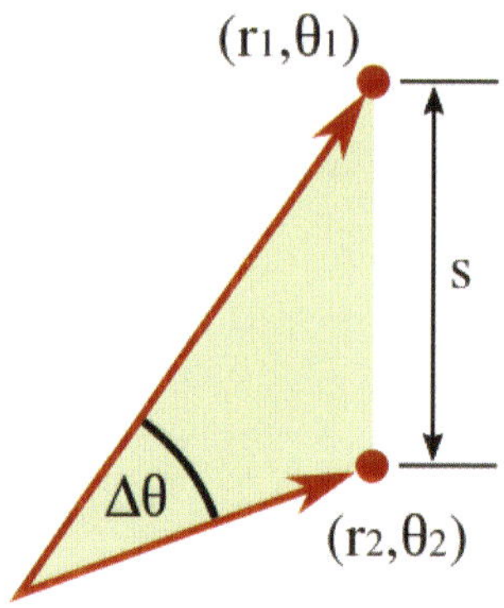

The separation s of two points (r_1, θ_1) and (r_2, θ_2) in polar coordinates is given by the law of cosines. Interior angle $\Delta\theta = \theta_1 - \theta_2$.

The Pythagorean theorem is a special case of the more general theorem relating the lengths of sides in any triangle, the law of cosines:

$$a^2 + b^2 - 2ab \cos\theta = c^2,$$

where θ is the angle between sides a and b.

When θ is 90 degrees ($\pi/2$ radians), then $\cos\theta = 0$, and the formula reduces to the usual Pythagorean theorem.

Arbitrary Triangle

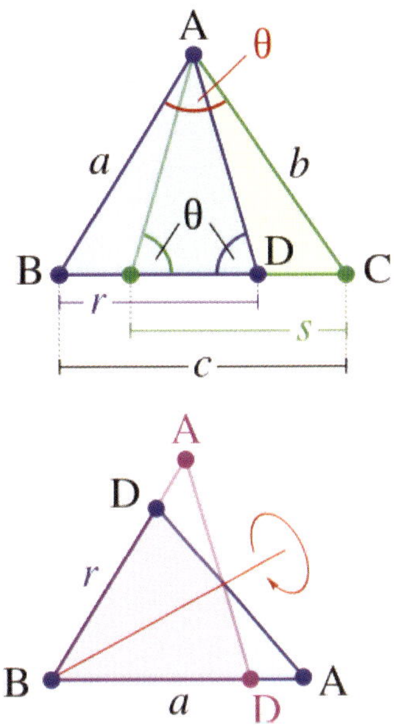

Generalization of Pythagoras's theorem by Tâbit ibn Qorra. Lower panel: reflection of triangle ABD (top) to form triangle DBA, similar to triangle ABC (top).

At any selected angle of a general triangle of sides *a, b,* c, inscribe an isosceles triangle such that the equal angles at its base θ are the same as the selected angle. Suppose the selected angle θ is opposite the side labeled c. Inscribing the isosceles triangle forms triangle *ABD* with angle θ opposite side *a* and with side *r* along c. A second triangle is formed with angle θ opposite side *b* and a side with length *s* along c, as shown in the figure. Thābit ibn Qurra stated that the sides of the three triangles were related as:

$$a^2 + b^2 = c(r + s).$$

As the angle θ approaches π/2, the base of the isosceles triangle narrows, and lengths *r* and *s* overlap less and less. When θ = π/2, *ADB* becomes a right triangle, *r* + *s* = c, and the original Pythagorean theorem is regained.

One proof observes that triangle *ABC* has the same angles as triangle *ABD*, but in opposite order. (The two triangles share the angle at vertex B, both contain the angle θ, and so also have the same third angle by the triangle postulate.) Consequently, *ABC* is similar to the reflection of *ABD*, the triangle *DBA* in the lower panel. Taking the ratio of sides opposite and adjacent to θ,

$$\frac{c}{a} = \frac{a}{r}.$$

Likewise, for the reflection of the other triangle,

$$\frac{c}{b} = \frac{b}{s}.$$

Clearing fractions and adding these two relations:

$$cr + cs = a^2 + b^2 ,$$

the required result.

The theorem remains valid if the angle θ is obtuse so the lengths r and s are non-overlapping.

General Triangles using Parallelograms

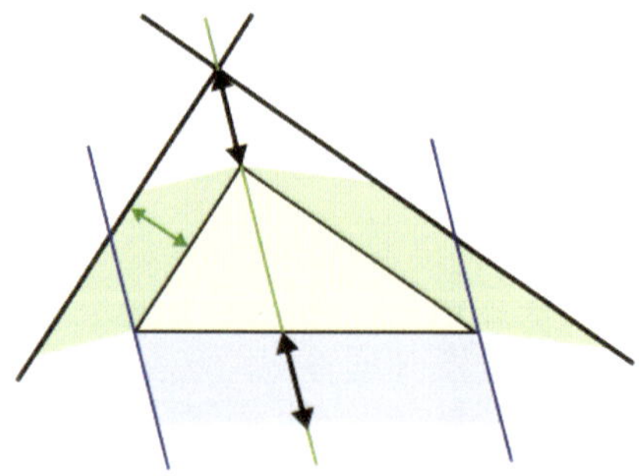

Generalization for arbitrary triangles, green area = blue area

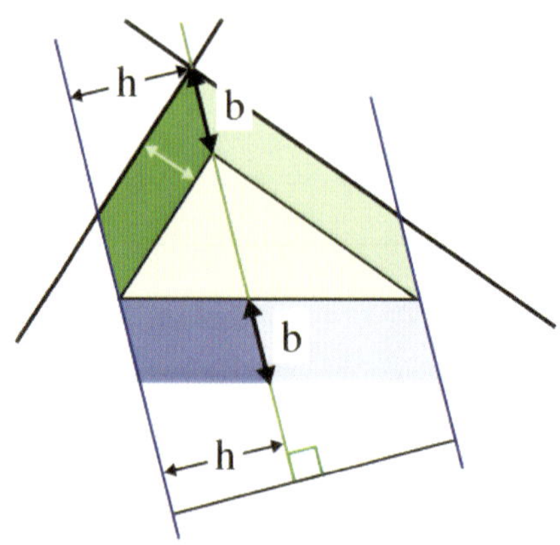

Construction for proof of parallelogram generalization

Pappus's area theorem is a further generalization, that applies to triangles that are not right triangles, using parallelograms on the three sides in place of squares (squares are a special case, of course). The upper figure shows that for a scalene triangle, the area of the parallelogram on the longest side is the sum of the areas of the parallelograms on the other two sides, provided the parallelogram on the long side is constructed as indicated (the dimensions labeled with arrows are the same, and determine the sides of the bottom parallelogram). This replacement of squares with parallelograms bears a clear resemblance to the original Pythagoras's theorem, and was considered a generalization by Pappus of Alexandria in 4 AD.

The lower figure shows the elements of the proof. Focus on the left side of the figure. The left green parallelogram has the same area as the left, blue portion of the bottom parallelogram because both have the same base b and height h. However, the left green parallelogram also has the same area as the left green parallelogram of the upper figure, because they have the same base (the upper left side of the triangle) and the same height normal to that side of the triangle. Repeating the argument for the right side of the figure, the bottom parallelogram has the same area as the sum of the two green parallelograms.

Solid Geometry

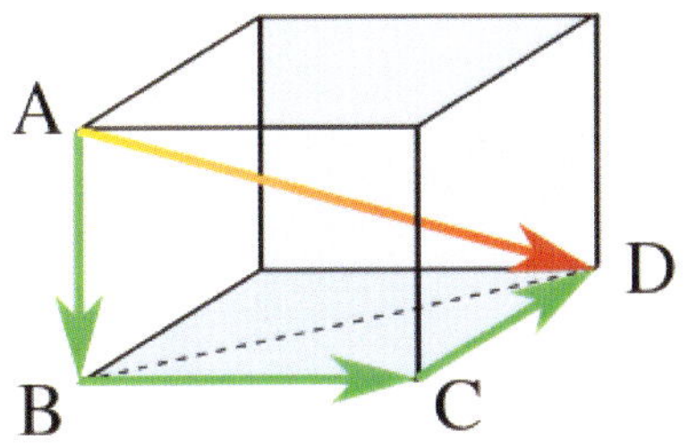

Pythagoras's theorem in three dimensions relates the diagonal AD to the three sides.

A tetrahedron with outward facing right-angle corner

In terms of solid geometry, Pythagoras's theorem can be applied to three dimensions as follows. Consider a rectangular solid as shown in the figure. The length of diagonal *BD* is found from Pythagoras's theorem as:

$$\overline{BD}^2 = \overline{BC}^2 + \overline{CD}^2$$

Where these three sides form a right triangle. Using horizontal diagonal *BD* and the vertical edge *AB*, the length of diagonal *AD* then is found by a second application of Pythagoras's theorem as:

$$\overline{AD}^2 = \overline{AB}^2 + \overline{BD}^2$$

or, doing it all in one step:

$$\overline{AD}^2 = \overline{AB}^2 + \overline{BC}^2 + \overline{CD}^2 .$$

This result is the three-dimensional expression for the magnitude of a vector v (the diagonal AD) in terms of its orthogonal components $\{ v_k \}$ (the three mutually perpendicular sides):

$$\| \mathbf{v} \|^2 = \sum_{k=1}^{3} \| \mathbf{v}_k \|^2 .$$

This one-step formulation may be viewed as a generalization of Pythagoras's theorem to higher dimensions. However, this result is really just the repeated application of the original Pythagoras's theorem to a succession of right triangles in a sequence of orthogonal planes.

A substantial generalization of the Pythagorean theorem to three dimensions is de Gua's theorem, named for Jean Paul de Gua de Malves: If a tetrahedron has a right angle corner (like a corner of a cube), then the square of the area of the face opposite the right angle corner is the sum of the squares of the areas of the other three faces. This result can be generalized as in the "n-dimensional Pythagorean theorem":

Let $x_1, x_2, \ldots, x_n$ be orthogonal vectors in R^n. Consider the n-dimensional simplex S with vertices $0, x_1, \ldots, x_n$. (Think of the $(n-1)$-dimensional simplex with vertices $x_1, \ldots, x_n$ not including the origin as the "hypotenuse" of S and the remaining $(n-1)$-dimensional faces of S as its "legs".) Then the square of the volume of the hypotenuse of S is the sum of the squares of the volumes of the n legs.

This statement is illustrated in three dimensions by the tetrahedron in the figure. The "hypotenuse" is the base of the tetrahedron at the back of the figure, and the "legs" are the three sides emanating from the vertex in the foreground. As the depth of the base from the vertex increases, the area of the "legs" increases, while that of the base is fixed. The theorem suggests that when this depth is at the value creating a right vertex, the generalization of Pythagoras's theorem applies. In a different wording:

Given an n-rectangular n-dimensional simplex, the square of the $(n-1)$-content of the facet opposing the right vertex will equal the sum of the squares of the $(n-1)$-contents of the remaining facets.

Inner Product Spaces

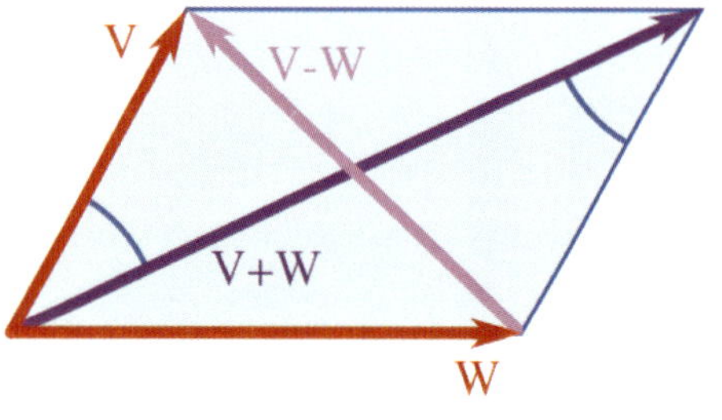

Vectors involved in the parallelogram law

The Pythagorean theorem can be generalized to inner product spaces, which are generalizations of the familiar 2-dimensional and 3-dimensional Euclidean spaces. For example, a function may be considered as a vector with infinitely many components in an inner product space, as in functional analysis.

In an inner product space, the concept of perpendicularity is replaced by the concept of orthogonality: two vectors v and w are orthogonal if their inner product $\langle \mathbf{v}, \mathbf{w} \rangle$ is zero. The inner product is a generalization of the dot product of vectors. The dot product is called the *standard* inner product or the *Euclidean* inner product. However, other inner products are possible.

The concept of length is replaced by the concept of the norm ||v|| of a vector v, defined as:

$$\| \mathbf{v} \| \equiv \sqrt{\langle \mathbf{v}, \mathbf{v} \rangle}.$$

In an inner-product space, the Pythagorean Theorem states that for any two orthogonal vectors v and w we have

$$\left\| \mathbf{v} + \mathbf{w} \right\|^2 = \left\| \mathbf{v} \right\|^2 + \left\| \mathbf{w} \right\|^2.$$

Here the vectors v and w are akin to the sides of a right triangle with hypotenuse given by the vector sum v + w. This form of the Pythagorean Theorem is a consequence of the properties of the inner product:

$$\left\| \mathbf{v} + \mathbf{w} \right\|^2 = \langle \mathbf{v} + \mathbf{w}, \mathbf{v} + \mathbf{w} \rangle = \langle \mathbf{v}, \mathbf{v} \rangle + \langle \mathbf{w}, \mathbf{w} \rangle + \langle \mathbf{v}, \mathbf{w} \rangle + \langle \mathbf{w}, \mathbf{v} \rangle = \left\| \mathbf{v} \right\|^2 + \left\| \mathbf{w} \right\|^2,$$

where the inner products of the cross terms are zero, because of orthogonally.

A further generalization of the Pythagorean theorem in an inner product space to non-orthogonal vectors is the *parallelogram law*:

$$2 \| \mathbf{v} \|^2 + 2 \| \mathbf{w} \|^2 = \| \mathbf{v} + \mathbf{w} \|^2 + \| \mathbf{v} - \mathbf{w} \|^2,$$

Which says that twice the sum of the squares of the lengths of the sides of a parallelogram is the sum of the squares of the lengths of the diagonals? Any norm that satisfies this equality is *ipso facto* a norm corresponding to an inner product.

The Pythagorean identity can be extended to sums of more than two orthogonal vectors. If v_1, v_2, ..., v_n are pairwise-orthogonal vectors in an inner-product space, then application of the Pythagorean theorem to successive pairs of these vectors results in the equation

$$\left\| \sum_{k=1}^{n} \mathbf{v}_k \right\|^2 = \sum_{k=1}^{n} \| \mathbf{v}_k \|^2$$

Sets of *m*-dimensional objects in *n*-dimensional Space

Another generalization of the Pythagorean theorem applies to Lebesgue-measurable sets of objects in any number of dimensions. Specifically, the square of the measure of an *m*-dimensional set of objects in one or more parallel *m*-dimensional flats in *n*-dimensional Euclidean space is equal to the sum of the squares of the measures of the orthogonal projections of the object(s) onto all *m*-dimensional coordinate subspaces.

In mathematical terms:

$$\mu_{ms}^2 = \sum_{i=1}^{x} \mathbf{i}^{\mathbf{2}}{}_{mp_i}$$

where:

- μ_m is a measure in m-dimensions (a length in one dimension, an area in two dimensions, a volume in three dimensions, etc.).
- s is a set of one or more non-overlapping m-dimensional objects in one or more parallel m-dimensional flats in n-dimensional Euclidean space.
- μ_{ms} is the total measure (sum) of the set of m-dimensional objects.
- p represents an m-dimensional projection of the original set onto an orthogonal coordinate subspace.
- μ_{mp_i} is the measure of the m-dimensional set projection onto m-dimensional coordinate subspace i. Because object projections can overlap on a coordinate subspace, the measure of each object projection in the set must be calculated individually, then measures of all projections added together to provide the total measure for the set of projections on the given coordinate subspace.
- x is the number of orthogonal, m-dimensional coordinate subspaces in n-dimensional space (R^n) onto which the m-dimensional objects are projected ($m \le n$):

$$= \left(\quad \right) = \frac{}{m!(n \quad m)!}$$

Non-Euclidean Geometry

The Pythagorean theorem is derived from the axioms of Euclidean geometry, and in fact, the Pythagorean theorem given above does not hold in a non-Euclidean geometry. (The Pythagorean theorem has been shown, in fact, to be equivalent to Euclid's Parallel (Fifth) Postulate.) In other words, in non-Euclidean geometry, the relation between the sides of a triangle must necessarily take a non-Pythagorean form. For example, in spherical geometry, all three sides of the right triangle (say a, b, and c) bounding an octant of the unit sphere have length equal to $\pi/2$, and all its angles are right angles, which violates the Pythagorean theorem because $a^2 + b^2 \neq c^2$.

Here two cases of non-Euclidean geometry are considered—spherical geometry and hyperbolic plane geometry; in each case, as in the Euclidean case for non-right triangles, the result replacing the Pythagorean theorem follows from the appropriate law of cosines.

However, the Pythagorean theorem remains true in hyperbolic geometry and elliptic geometry if the condition that the triangle be right is replaced with the condition that two of the angles sum to the third, say $A+B = C$. The sides are then related as follows: the sum of the areas of the circles with diameters a and b equals the area of the circle with diameter c.

Spherical Geometry

For any right triangle on a sphere of radius R (for example, if γ in the figure is a right angle), with sides a, b, c, the relation between the sides takes the form:

$$\cos\left(\frac{c}{R}\right) = \cos\left(\frac{a}{R}\right)\cos\left(\frac{b}{R}\right).$$

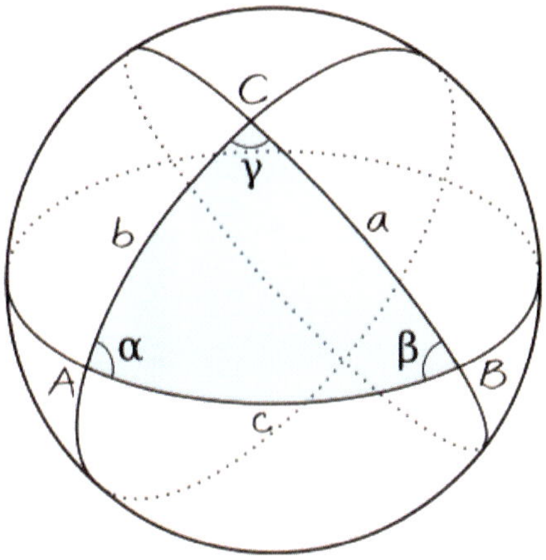

Spherical triangle

This equation can be derived as a special case of the spherical law of cosines that applies to all spherical triangles:

$$\cos\left(\frac{c}{R}\right)=\cos\left(\frac{a}{R}\right)\cos\left(\frac{b}{R}\right)+\sin\left(\frac{a}{R}\right)\sin\left(\frac{b}{R}\right)\cos\gamma .$$

By expressing the Maclaurin series for the cosine function as an asymptotic expansion with the remainder term in big O notation,

$$\cos x=1-\frac{x^2}{2}+O(x^4) \text{ as } x\to 0,$$

It can be shown that as the radius R approaches infinity and the arguments a/R, b/R, and c/R tend to zero, the spherical relation between the sides of a right triangle approaches the Euclidean form of the Pythagorean theorem. Substituting the asymptotic expansion for each of the cosines into the spherical relation for a right triangle yields

$$1-\frac{1}{2}\left(\frac{c}{R}\right)^2+O\left(\frac{1}{R^4}\right)=\left[1-\frac{1}{2}\left(\frac{a}{R}\right)^2+O\left(\frac{1}{R^4}\right)\right]\left[1-\frac{1}{2}\left(\frac{b}{R}\right)^2+O\left(\frac{1}{R^4}\right)\right] \text{ as } R\to\infty .$$

The constants a^4, b^4, and c^4 have been absorbed into the big O remainder terms since they are independent of the radius R. This asymptotic relationship can be further simplified by multiplying out the bracketed quantities, cancelling the ones, multiplying through by −2, and collecting all the error terms together:

$$\left(\frac{c}{R}\right)^2=\left(\frac{a}{R}\right)^2+\left(\frac{b}{R}\right)^2+O\left(\frac{1}{R^4}\right) \text{ as } R\to\infty .$$

After multiplying through by R^2, the Euclidean Pythagorean relationship $c^2 = a^2 + b^2$ is recovered in the limit as the radius R approaches infinity (since the remainder term tends to zero):

$$c^2=a^2+b^2+O\left(\frac{1}{R^2}\right) \text{ as } R\to\infty .$$

For small right triangles (a, b << R), the cosines can be eliminated to avoid loss of significance, giving

$$\sin^2\frac{c}{2R}=\sin^2\frac{a}{2R}+\sin^2\frac{b}{2R}-2\sin^2\frac{a}{2R}\sin^2\frac{b}{2R}.$$

Hyperbolic Geometry

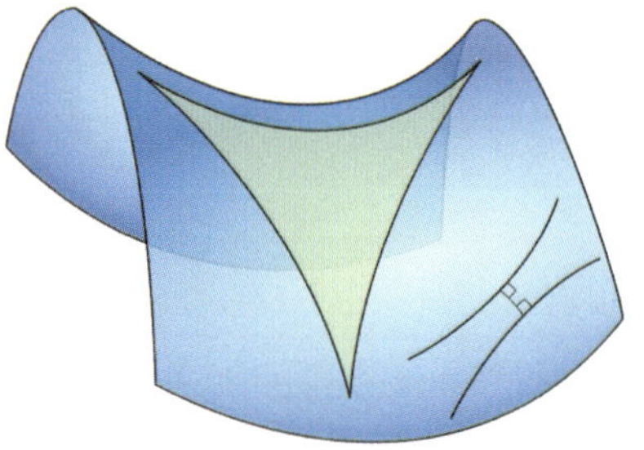

Hyperbolic triangle

In a hyperbolic space with uniform curvature $-1/R^2$, for a right triangle with legs a, b, and hypotenuse c, the relation between the sides takes the form:

$$\cosh\frac{c}{R}=\cosh\frac{a}{R}\cosh\frac{b}{R}$$

where cosh is the hyperbolic cosine. This formula is a special form of the hyperbolic law of cosines that applies to all hyperbolic triangles:

$$\cosh\frac{c}{R}=\cosh\frac{a}{R}\cosh\frac{b}{R}-\sinh\frac{a}{R}\sinh\frac{b}{R}\cos\gamma,$$

with γ the angle at the vertex opposite the side c.

By using the Maclaurin series for the hyperbolic cosine, $\cosh x \approx 1 + x^2/2$, it can be shown that as a hyperbolic triangle becomes very small (that is, as a, b, and c all approach zero), the hyperbolic relation for a right triangle approaches the form of Pythagoras's theorem.

For small right triangles (a, b << R), the hyperbolic cosines can be eliminated to avoid loss of significance, giving

$$\sinh^2\frac{c}{2R}=\sinh^2\frac{a}{2R}+\sinh^2\frac{b}{2R}+2\sinh^2\frac{a}{2R}\sinh^2\frac{b}{2R}.$$

Very Small Triangles

For any uniform curvature K (positive, zero, or negative), in very small right triangles ($|K|a^2$, $|K|b^2$ << 1) with hypotenuse c, it can be shown that

$$c^2=a^2+b^2-\frac{K}{3}a^2b^2-\frac{K^2}{45}a^2b^2(a^2+b^2)-\frac{2K^3}{945}a^2b^2(a^2-b^2)^2+O(K^4c^{10}).$$

Differential Geometry

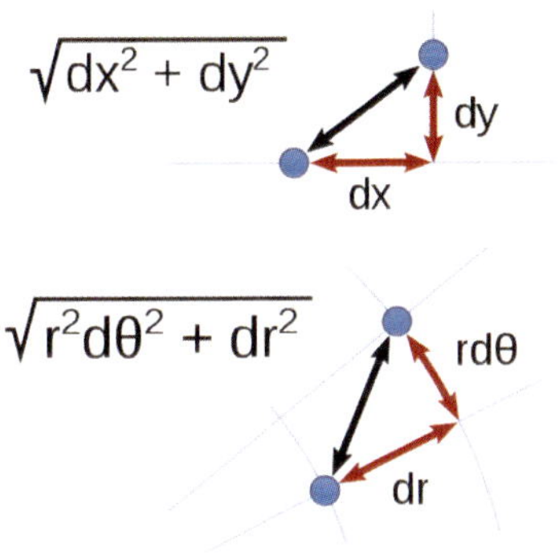

Distance between infinitesimally separated points in Cartesian coordinates (top) and polar coordinates (bottom), as given by Pythagoras's theorem

On an infinitesimal level, in three dimensional space, Pythagoras's theorem describes the distance between two infinitesimally separated points as:

$$ds^2 = dx^2 + dy^2 + dz^2,$$

with *ds* the element of distance and (*dx*, *dy*, *dz*) the components of the vector separating the two points. Such a space is called a Euclidean space. However, in Riemannian geometry, a generalization of this expression useful for general coordinates (not just Cartesian) and general spaces (not just Euclidean) takes the form:

$$ds^2 = \sum_{i,j}^{n} g_{ij}\, dx_i\, dx_j$$

Which is called the metric tensor. (Sometimes, by abuse of language, the same term is applied to the set of coefficients g_{ij}.) It may be a function of position, and often describes curved space. A simple example is Euclidean (flat) space expressed in curvilinear coordinates. For example, in polar coordinates:

$$ds^2 = dr^2 + r^2 d\theta^2 .$$

Pythagorean Triple

Three integers a, b, and c that satisfy $a^2 + b^2 = c^2$ are called *Pythagorean Triples*. There are infinitely many such numbers and there also exists a way to generate all the triples. Let n and m be integers, n>m. Then define

$$(*)\ \ a = n^2 - m^2, b = 2nm, c = n^2 + m^2$$

The three number a, b, and c always form a Pythagorean triple. The proof is simple:

$$\begin{aligned}(n^2 - m^2)^2 + (2mn)^2 &= n^4 - 2n^2m^2 + m^4 + 4n^2m^2 \\ &= n^4 + 2n^2m^2 + m^4 \\ &= (n^2 + m^2)^2\end{aligned}$$

The formulas were known to Euclid and used by Diophantus to obtain Pythagorean triples with special properties.

Some examples of Pythagorean triples:

$$\begin{matrix} (3,4,5) & (5,12,13) & (6,8,10) & (7,24,25) \\ (8,15,17) & (9,12,15) & (9,40,41) & (10,24,26) \\ (11,60,61) & (12,16,20) & (12,35,37) & (13,84,85) \end{matrix}$$

Some Pythagorean triples.

The triples written in red are multiples of each other and so are the triples written in blue: you get $(6,8,10)$, $(9,12,15)$ and $(12,16,20)$ by multiplying the components of $(3,4,5)$ by 2, 3 and 4 respectively, and you get $(10,24,26)$ by multiplying the components of $(5,12,13)$ by 2.

In general, if k is a positive whole number and (a,b,c) is a Pythagorean triple, then so is (ka, kb, kc), since

$$(ka)^2+(kb)^2=k^2a^2+k^2b^2=k^2\left(a^2+b^2\right)=k^2c^2=(kc)^2$$

Geometrically, if one Pythagorean triple is a multiple of another, then the corresponding triangles are similar.

If a Pythagorean triple isn't a multiple of another Pythagorean triple, then we say that it is a *primitive triple*. You can recognise a primitive Pythagorean triple by the fact that the numbers a and b do not have a common divisor. In our example $(3,4,5)$ is a primitive Pythagorean triple while $(6,8,10)$ $(9,12,15)$ and $(12,16,20)$ are not. Similarly $(5,12,13)$ is a primitive triple whereas $(10,24,26)$ is not.

Pythagoras as depicted by Raffaello Sanzio in his painting The school of Athens.

Elementary properties of primitive Pythagorean triples

General Properties

The properties of a primitive Pythagorean triple (a, b, c) with $a < b < c$ (without specifying which of a or b is even and which is odd) include:

- $\frac{(c-a)(c-b)}{2}$ is always a perfect square. As it is only a necessary condition but not a sufficient one, it can be used in checking if a given triple of numbers is *not* a Pythagorean triple when they fail the test. For example, the triple {6, 12, 18} passes the test that $(c - a)(c - b)/2$ is a perfect square, but it is not a Pythagorean triple.
- When a triple of numbers a, b and c forms a primitive Pythagorean triple, then (c minus the even leg) and one-half of (c minus the odd leg) are both perfect squares; however this is not a sufficient condition, as the numbers {1, 8, 9} pass the perfect squares test but are not a Pythagorean triple since $1^2 + 8^2 \neq 9^2$.
- At most one of a, b, c is a square.
- The area of a Pythagorean triangle cannot be the square or twice the square of a natural number.
- Exactly one of a, b is odd; c is odd.
- Exactly one of a, b is divisible by 3.
- Exactly one of a, b is divisible by 4.
- Exactly one of a, b, c is divisible by 5.
- The largest number that always divides abc is 60.
- All prime factors of c are primes of the form $4n + 1$. Therefore c is of the form $4n + 1$.
- The area ($K = ab/2$) is a congruent number divisible by 6.
- In every Pythagorean triple, the radius of the incircle and the radii of the three excircles are natural numbers. Specifically, for a primitive triple the radius of the incircle is $r = n(m - n)$, and the radii of the excircles opposite the sides $m^2 - n^2$, $2mn$, and the hypotenuse $m^2 + n^2$ are respectively $m(m - n)$, $n(m + n)$, and $m(m + n)$.
- As for any right triangle, the converse of Thales' theorem says that the diameter of the circumcircle equals the hypotenuse; hence for primitive triples the circumdiameter is $m^2 + n^2$, and the circumradius is half of this and thus is rational but non-integer (since m and n have opposite parity).
- When the area of a Pythagorean triangle is multiplied by the curvatures of its incircle and 3 excircles, the result is four positive integers $w > x > y > z$, respectively. Integers $-w, x, y, z$ satisfy Descartes's Circle Equation. Equivalently, the radius of the outer Soddy circle of any right triangle is equal to its semiperimeter. The outer Soddy center is located at D, where $ACBD$ is a rectangle, ACB the right triangle and AB its hypotenuse.
- There are no Pythagorean triples in which the hypotenuse and one leg are the legs of

another Pythagorean triple; this is one of the equivalent forms of Fermat's right triangle theorem.

- Each primitive Pythagorean triangle has a ratio of area to squared semiperimeter that is unique to itself and is given by

$$\frac{K}{s^2} = \frac{n(m-n)}{m(m+n)} = 1 - \frac{c}{s}.$$

- No primitive Pythagorean triangle has an integer altitude from the hypotenuse; that is, every primitive Pythagorean triangle is indecomposable.
- The set of all primitive Pythagorean triples forms a rooted ternary tree in a natural way.
- Neither of the acute angles of a Pythagorean triangle can be a rational number of degrees. (This follows from Niven's theorem.)

Special Cases

In addition, special Pythagorean triples with certain additional properties can be guaranteed to exist:

- Every integer greater than 2 that is not congruent to 2 mod 4 (in other words, every integer greater than 2 which is *not* of the form $4k + 2$) is part of a primitive Pythagorean triple. (If the integer has the form $4k$, one may take $n = 1$ and $m = 2k$ in Euler's formula; if the integer is $2k + 1$, one may take $n = k$ and $m = k + 1$.)
- Every integer greater than 2 is part of a primitive or non-primitive Pythagorean triple. For example, the integers 6, 10, 14, and 18 are not part of primitive triples, but are part of the non-primitive triples (6, 8, 10), (14, 48, 50) and (18, 80, 82).
- There exist infinitely many Pythagorean triples in which the hypotenuse and the longest leg differ by exactly one. Such triples are necessarily primitive and have the form $(2n + 1, 2n^2 + 2n, 2n^2 + 2n + 1)$. This results from Euler's formula by remarking that the condition implies that the triple is primitive and must verify $(m^2 + m^2) - 2mn = 1$. This implies $(m - n)^2 = 1$, and thus $m = n + 1$. The above form of the triples results thus of substituting m for $n + 1$ in Euler's formula.
- There exist infinitely many primitive Pythagorean triples in which the hypotenuse and the longest leg differ by exactly two. They are all primitive, and are obtained by putting $n = 1$ in Euler's formula. More generally, for every integer $k > 0$, there exist infinitely many primitive Pythagorean triples in which the hypotenuse and the odd leg differ by $2k^2$. They are obtained by putting $n = k$ in Euler's formula.
- There exist infinitely many Pythagorean triples in which the two legs differ by exactly one. For example, $20^2 + 21^2 = 29^2$; these are generated by Euclid's formula when $\frac{m-n}{n}$ is a convergent to $\sqrt{2}$.
- For each natural number n, there exist n Pythagorean triples with different hypotenuses and the same area.
- For each natural number n, there exist at least n different Pythagorean triples with the same leg a, where a is some natural number.

- For each natural number n, there exist at least n different Pythagorean triples with the same hypotenuse.
- There exist infinitely many Pythagorean triples with square numbers for both the hypotenuse c and the sum of the legs $a + b$. According to Fermat, the smallest such triple has sides $a = 4{,}565{,}486{,}027{,}761$; $b = 1{,}061{,}652{,}293{,}520$; and $c = 4{,}687{,}298{,}610{,}289$. Here $a + b = 2{,}372{,}159^2$ and $c = 2{,}165{,}017^2$. This is generated by Euclid's formula with parameter values $m = 2{,}150{,}905$ and $n = 246{,}792$.
- There exist non-primitive Pythagorean triangles with integer altitude from the hypotenuse. Such Pythagorean triangles are known as decomposable since they can be split along this altitude into two separate and smaller Pythagorean triangles.

Stereographic Approach

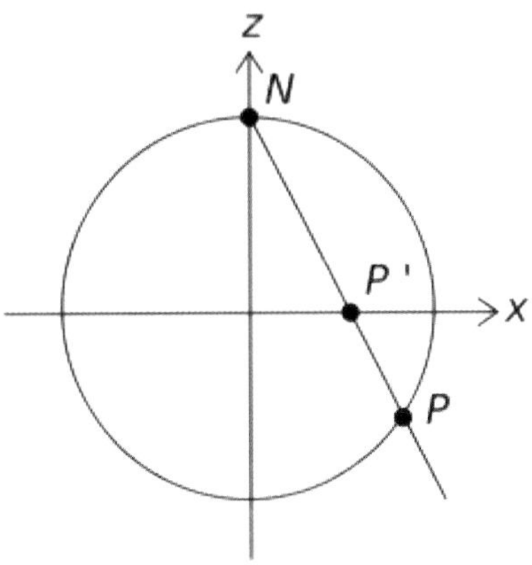

Stereographic projection of the unit circle onto the x-axis. Given a point P on the unit circle, draw a line from P to the point $N = (0, 1)$ (the *north pole*).

The point P' where the line intersects the x-axis is the stereographic projection of P. Inversely, starting with a point P' on the x-axis, and drawing a line from P' to N, the inverse stereographic projection is the point P where the line intersects the unit circle.

There is a correspondence between points on the unit circle with rational coordinates and primitive Pythagorean triples. At this point, Euclid's formulae can be derived either by methods of trigonometry or equivalently by using the stereographic projection.

For the stereographic approach, suppose that P' is a point on the x-axis with rational coordinates

$$P' = \left(\frac{m}{n}, 0\right).$$

Then, it can be shown by basic algebra that the point P has coordinates

$$P = \left(\frac{2\left(\frac{m}{n}\right)}{\left(\frac{m}{n}\right)^2 + 1}, \frac{\left(\frac{m}{n}\right)^2 - 1}{\left(\frac{m}{n}\right)^2 + 1}\right) = \left(\frac{2mn}{m^2 + n^2}, \frac{m^2 - n^2}{m^2 + n^2}\right)$$

This establishes that each rational point of the x-axis goes over to a rational point of the unit circle. The converse, that every rational point of the unit circle comes from such a point of the x-axis, follows by applying the inverse stereographic projection. Suppose that $P(x, y)$ is a point of the unit circle with x and y rational numbers. Then the point P' obtained by stereographic projection onto the x-axis has coordinates

$$\left(\frac{\quad}{\quad},0\right)$$

which is rational.

In terms of algebraic geometry, the algebraic variety of rational points on the unit circle is birational to the affine line over the rational numbers. The unit circle is thus called a rational curve, and it is this fact which enables an explicit parameterization of the (rational number) points on it by means of rational functions.

Pythagorean Triangles in a 2D Lattice

A 2D lattice is a regular array of isolated points where if any one point is chosen as the Cartesian origin (0, 0), then all the other points are at (x, y) where x and y range over all positive and negative integers. Any Pythagorean triangle with triple (a, b, c) can be drawn within a 2D lattice with vertices at coordinates (0, 0), $(a, 0)$ and $(0, b)$. The count of lattice points lying strictly within the bounds of the triangle is given by $\frac{(a-1)(b-1)-\gcd(a,b)+1}{2}$ for primitive Pythagorean triples this interior lattice count is $\frac{(a-1)(b-1)}{2}$ The area (by Pick's theorem equal to one less than the interior lattice count plus half the boundary lattice count) equals $\frac{ab}{2}$.

The first occurrence of two primitive Pythagorean triples sharing the same area occurs with triangles with sides (20, 21, 29), (12, 35, 37) and common area 210 (sequence A093536 in the OEIS). The first occurrence of two primitive Pythagorean triples sharing the same interior lattice count occurs with (18108, 252685, 253333), (28077, 162964, 165365) and interior lattice count 2287674594 (sequence A225760 in the OEIS). Three primitive Pythagorean triples have been found sharing the same area: (4485, 5852, 7373), (3059, 8580, 9109), (1380, 19019, 19069) with area 13123110. As yet, no set of three primitive Pythagorean triples have been found sharing the same interior lattice count.

Enumeration of Primitive Pythagorean Triples

By Euclid's formula all primitive Pythagorean triples can be generated from integers m and n with $m > n > 0$, $m+n$ odd and $\gcd(m,n) = 1$. Hence there is a 1 to 1 mapping of rationals (in lowest terms) to primitive Pythagorean triples where $\frac{n}{m}$ is in the interval $(0,1)$ and $m+n$ odd.

The reverse mapping from a primitive triple (a,b,c) where $c > b > a > 0$ to a rational $\frac{n}{m}$ is achieved by studying the two sums $a+c$ and $b+c$. One of these sums will be a square that can be equated to $(m+n)^2$ and the other will be twice a square that can be equated to $2m^2$. It is then possible to determine the rational $\frac{n}{m}$.

In order to enumerate primitive Pythagorean triples the rational can be expressed as an ordered pair (n,m) and mapped to an integer using a pairing function such as Cantor's pairing function. An example can be seen at (sequence A277557 in the OEIS). It begins

8,18,19,32,33,34,... and gives rationals

$\frac{1}{2}, \frac{2}{3}, \frac{1}{4}, \frac{3}{4}, \frac{2}{5}, \frac{1}{6}, \ldots$ these, in turn, generate primitive triples

(3,4,5),(5,12,13),(8,15,17),(7,24,25),(20,21,29),(12,35,37),...

Spinors and the Modular Group

Pythagorean triples can likewise be encoded into a matrix of the form

$$X = \begin{bmatrix} c+b & a \\ a & c-b \end{bmatrix}.$$

A matrix of this form is symmetric. Furthermore, the determinant of X is

$$\det X = c^2 - a^2 - b^2$$

which is zero precisely when (a,b,c) is a Pythagorean triple. If X corresponds to a Pythagorean triple, then as a matrix it must have rank 1.

Since X is symmetric, it follows from a result in linear algebra that there is a column vector $\xi = [m\ n]^{\mathrm{T}}$ such that the outer product

$$X = 2\begin{bmatrix} m \\ n \end{bmatrix}[m\ n] = 2\xi\xi^T$$

holds, where the T denotes the matrix transpose. The vector ξ is called a spinor (for the Lorentz group SO(1, 2)). In abstract terms, the Euclid formula means that each primitive Pythagorean triple can be written as the outer product with itself of a spinor with integer entries.

The modular group Γ is the set of 2×2 matrices with integer entries

$$A = \begin{bmatrix} \alpha & \beta \\ \gamma & \delta \end{bmatrix}$$

with determinant equal to one: $\alpha\delta - \beta\gamma = 1$. This set forms a group, since the inverse of a matrix in Γ is again in Γ, as is the product of two matrices in Γ. The modular group acts on the collection of all integer spinors. Furthermore, the group is transitive on the collection of integer spinors with relatively prime entries. For if $[m\ n]^{\mathrm{T}}$ has relatively prime entries, then

$$\begin{bmatrix} m & -v \\ n & u \end{bmatrix}\begin{bmatrix} 1 \\ 0 \end{bmatrix} = \begin{bmatrix} m \\ n \end{bmatrix}$$

where u and v are selected (by the Euclidean algorithm) so that $mu + nv = 1$.

By acting on the spinor ξ, the action of Γ goes over to an action on Pythagorean triples, provided one allows for triples with possibly negative components. Thus if A is a matrix in Γ, then

$$2(A\xi)(A\xi)^T = AXA^T$$

gives rise to an action on the matrix X. This does not give a well-defined action on primitive triples, since it may take a primitive triple to an imprimitive one. It is convenient at this point (per Trautman 1998) to call a triple (a,b,c) standard if $c > 0$ and either (a,b,c) are relatively prime or $(a/2,b/2,c/2)$ are relatively prime with $a/2$ odd. If the spinor $[m\ n]^T$ has relatively prime entries, then the associated triple (a,b,c) determined is a standard triple. It follows that the action of the modular group is transitive on the set of standard triples.

Alternatively, restrict attention to those values of m and n for which m is odd and n is even. Let the subgroup Γ of Γ be the kernel of the group homomorphism

$$\Gamma = \mathrm{SL}(2,\mathbf{Z}) \rightarrow \mathrm{SL}(2,\mathbf{Z}_2)$$

where $\mathrm{SL}(2,\mathrm{Z}_2)$ is the special linear group over the finite field Z_2 of integers modulo 2. Then Γ is the group of unimodular transformations which preserve the parity of each entry. Thus if the first entry of ξ is odd and the second entry is even, then the same is true of $A\xi$ for all $A \in \Gamma$. In fact, under the action, the group Γ acts transitively on the collection of primitive Pythagorean triples.

The group Γ is the free group whose generators are the matrices

$$U = \begin{bmatrix} 1 & 2 \\ 0 & 1 \end{bmatrix}, \qquad L = \begin{bmatrix} 1 & 0 \\ 2 & 1 \end{bmatrix}.$$

Consequently, every primitive Pythagorean triple can be obtained in a unique way as a product of copies of the matrices U and L.

Parent/child Relationships

By a result of Berggren, all primitive Pythagorean triples can be generated from the (3, 4, 5) triangle by using the three linear transformations T_1, T_2, T_3 below, where a, b, c are sides of a triple:

	new side a	new side b	new side c
T_1:	$a - 2b + 2c$	$2a - b + 2c$	$2a - 2b + 3c$
T_2:	$a + 2b + 2c$	$2a + b + 2c$	$2a + 2b + 3c$
T_3:	$-a + 2b + 2c$	$-2a + b + 2c$	$-2a + 2b + 3c$

In other words, every primitive triple will be a "parent" to three additional primitive triples. Starting from the initial node with $a = 3$, $b = 4$, and $c = 5$, the operation T_1 produces the new triple

$$(3 - (2\times 4) + (2\times 5), (2\times 3) - 4 + (2\times 5), (2\times 3) - (2\times 4) + (3\times 5)) = (5, 12, 13),$$

and similarly T_2 and T_3 produce the triples (21, 20, 29) and (15, 8, 17).

The linear transformations T_1, T_2, and T_3 have a geometric interpretation in the language of quadratic forms. They are closely related to (but are not equal to) reflections generating the orthogonal group of $x^2 + y^2 - z^2$ over the integers.

Relation to Gaussian Integers

Alternatively, Euclid's formulae can be analyzed and proven using the Gaussian integers. Gaussian integers are complex numbers of the form $\alpha = u + vi$, where u and v are ordinary integers and i is the square root of negative one. The units of Gaussian integers are ±1 and ±i. The ordinary integers are called the rational integers and denoted as Z. The Gaussian integers are denoted as Z[i]. The right-hand side of the Pythagorean theorem may be factored in Gaussian integers:

$$c^2 = a^2 + b^2 = (a+bi)\overline{(a+bi)} = (a+bi)(a-bi).$$

A primitive Pythagorean triple is one in which a and b are coprime, i.e., they share no prime factors in the integers. For such a triple, either a or b is even, and the other is odd; from this, it follows that c is also odd.

The two factors $z := a + bi$ and $z^* := a - bi$ of a primitive Pythagorean triple each equal the square of a Gaussian integer. This can be proved using the property that every Gaussian integer can be factored uniquely into Gaussian primes up to units. (This unique factorization follows from the fact that, roughly speaking, a version of the Euclidean algorithm can be defined on them.) The proof has three steps. First, if a and b share no prime factors in the integers, then they also share no prime factors in the Gaussian integers. (Assume $a = gu$ and $b = gv$ with Gaussian integers g, u and v and g not a unit. Then u and v lie on the same line through the origin. All Gaussian integers on such a line are integer multiples of some Gaussian integer h. But then the integer $gh \neq \pm 1$ divides both a and b.) Second, it follows that z and z^* likewise share no prime factors in the Gaussian integers. For if they did, then their common divisor δ would also divide $z + z^* = 2a$ and $z - z^* = 2ib$. Since a and b are coprime, that implies that δ divides $2 = (1 + i)(1 - i) = i(1 - i)^2$. From the formula $c^2 = zz^*$, that in turn would imply that c is even, contrary to the hypothesis of a primitive Pythagorean triple. Third, since c^2 is a square, every Gaussian prime in its factorization is doubled, i.e., appears an even number of times. Since z and z^* share no prime factors, this doubling is also true for them. Hence, z and z^* are squares.

Thus, the first factor can be written

$$a + bi = \varepsilon(m + ni)^2, \quad \varepsilon \in \{\pm 1, \pm i\}.$$

The real and imaginary parts of this equation give the two formulas:

$$\begin{cases} \varepsilon = +1, & a = +(m^2 - n^2), \quad b = +2mn; \\ \varepsilon = -1, & a = -(m^2 - n^2), \quad b = -2mn; \\ \varepsilon = +i, & a = -2mn, \quad b = +(m^2 - n^2); \\ \varepsilon = -i, & a = +2mn, \quad b = -(m^2 - n^2). \end{cases}$$

For any primitive Pythagorean triple, there must be integers m and n such that these two equations are satisfied. Hence, every Pythagorean triple can be generated from some choice of these integers.

As perfect Square Gaussian Integers

If we consider the square of a Gaussian integer we get the following direct interpretation of Euclid's formulae as representing a perfect square Gaussian integers.

$$(m+ni)^2 = (m^2 - n^2) + 2mni$$

Using the facts that the Gaussian integers are a Euclidean domain and that for a Gaussian integer p $|p|^2$ is always a square it is possible to show that a Pythagorean triples correspond to the square of a prime Gaussian integer if the hypotenuse is prime.

If the Gaussian integer is not prime then it is the product of two Gaussian integers p and q with $|p|^2$ and $|q|^2$ integers. Since magnitudes multiply in the Gaussian integers, the product must be $|p||q|$, which when squared to find a Pythagorean triple must be composite. The contrapositive completes the proof.

Relation to Ellipses with Integral Dimensions

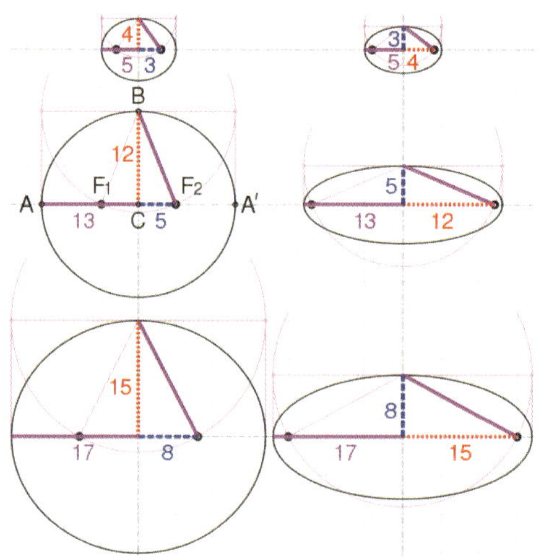

Relationship between Pythagorean triples and ellipses with integral linear eccentricity, and major and minor axes, for the first 3 Pythagorean triples

With reference to the figure and the definition of the foci of an ellipse, F_1 and F_2, for any point P on the ellipse, $F_1P + PF_2$ is constant.

As points A and B are both on the ellipse, $F_1A + AF_2 = F_1B + BF_2$. Due to symmetry, $F_1A + AF_2 = F_2A' + AF_2 = AA' = 2\ AC$, and $F_1B + BF_2 = 2\ BF_2$. Hence, $AC = BF_2$.

Thus, if BCF_2 is a right-angle triangle with integral sides, the separation of the foci, linear eccentricity, minor axis and major axis are all also integers.

Distribution of Triples

There are a number of results on the distribution of Pythagorean triples. In the scatter plot, a number of obvious patterns are already apparent. Whenever the legs (a, b) of a primitive triple appear in the plot, all integer multiples of (a, b) must also appear in the plot, and this property produces the appearance of lines radiating from the origin in the diagram.

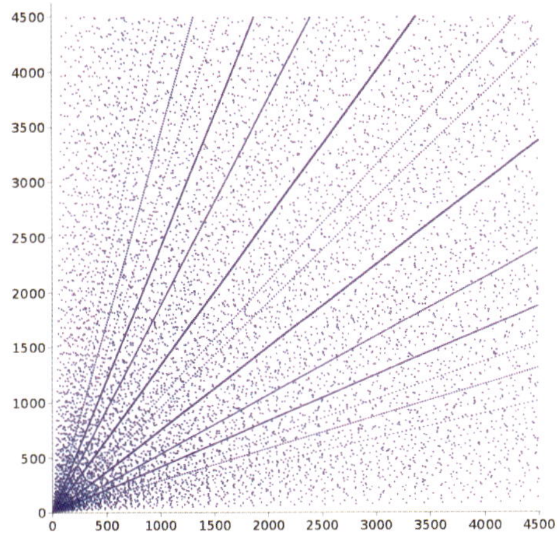

A scatter plot of the legs (a,b) of the first Pythagorean triples with a and b less than 4500.

Within the scatter, there are sets of parabolic patterns with a high density of points and all their foci at the origin, opening up in all four directions. Different parabolas intersect at the axes and appear to reflect off the axis with an incidence angle of 45 degrees, with a third parabola entering in a perpendicular fashion. Within this quadrant, each arc centered on the origin shows that section of the parabola that lies between its tip and its intersection with its semi-latus rectum.

These patterns can be explained as follows. If $a^2/4n$ is an integer, then $(a, |n-a^2/4n|, n+a^2/4n)$ is a Pythagorean triple. (In fact every Pythagorean triple (a, b, c) can be written in this way with integer n, possibly after exchanging a and b, since $n=(b+c)/2$ and a and b cannot both be odd.) The Pythagorean triples thus lie on curves given by $b=|n-a^2/4n|$, that is, parabolas reflected at the a-axis, and the corresponding curves with a and b interchanged. If a is varied for a given n (i.e. on a given parabola), integer values of b occur relatively frequently if n is a square or a small multiple of a square. If several such values happen to lie close together, the corresponding parabolas approximately coincide, and the triples cluster in a narrow parabolic strip. For instance, $38^2 = 1444$, $2 \times 27^2 = 1458$, $3 \times 22^2 = 1452$, $5 \times 17^2 = 1445$ and $10 \times 12^2 = 1440$; the corresponding parabolic strip around $n \approx 1450$ is clearly visible in the scatter plot.

The angular properties described above follow immediately from the functional form of the parabolas. The parabolas are reflected at the a-axis at $a = 2n$, and the derivative of b with respect to a at this point is -1; hence the incidence angle is 45°. Since the clusters, like all triples, are repeated at integer multiples, the value $2n$ also corresponds to a cluster. The corresponding parabola intersects the b-axis at right angles at $b = 2n$, and hence its reflection upon interchange of a and b intersects the a-axis at right angles at $a = 2n$, precisely where the parabola for n is reflected at the a-axis. (The same is of course true for a and b interchanged.)

Albert Fässler and others provide insights into the significance of these parabolas in the context of conformal mappings.

Special cases and Related Equations

The Platonic Sequence

The case $n = 1$ of the more general construction of Pythagorean triples has been known for a long time. Proclus, in his commentary to the 47th Proposition of the first book of Euclid's Elements, describes it as follows:

Certain methods for the discovery of triangles of this kind are handed down, one which they refer to Plato, and another to Pythagoras. (The latter) starts from odd numbers. For it makes the odd number the smaller of the sides about the right angle; then it takes the square of it, subtracts unity and makes half the difference the greater of the sides about the right angle; lastly it adds unity to this and so forms the remaining side, the hypotenuse. For the method of Plato argues from even numbers. It takes the given even number and makes it one of the sides about the right angle; then, bisecting this number and squaring the half, it adds unity to the square to form the hypotenuse, and subtracts unity from the square to form the other side about the right angle. Thus it has formed the same triangle that which was obtained by the other method.

In equation form, this becomes:

a is odd (Pythagoras):

$$\text{side } a : \text{side } b = \frac{a^2-1}{2} : \text{side } c = \frac{a^2+1}{2}.$$

a is even (Plato, c. 380 BC):

$$\text{side } a : \text{side } b = \left(\frac{a}{2}\right)^2 - 1 : \text{side } c = \left(\frac{a}{2}\right)^2 + 1$$

It can be shown that all Pythagorean triples can be obtained, with appropriate rescaling, from the basic Platonic sequence (a, $(a^2 - 1)/2$ and $(a^2 + 1)/2$) by allowing a to take non-integer rational values. If a is replaced with the fraction m/n in the sequence, the result is equal to the 'standard' triple generator ($2mn$, $m^2 - n^2$,$m^2 + n^2$) after rescaling. It follows that every triple has a corresponding rational a value which can be used to generate a similar triangle (one with the same three angles and with sides in the same proportions as the original). For example, the Platonic equivalent of (56, 33, 65) is generated by $a = m/n = 7/4$ as $(a, (a^2 -1)/2, (a^2+1)/2) = (56/32, 33/32, 65/32)$. The Platonic sequence itself can be derived by following the steps for 'splitting the square' described in Diophantus II.VIII.

The Jacobi–Madden Equation

The equation,

$$a^4 + b^4 + c^4 + d^4 = (a+b+c+d)^4$$

is equivalent to the special Pythagorean triple,

$$(a^2 + ab + b^2)^2 + (c^2 + cd + d^2)^2 = ((a+b)^2 + (a+b)(c+d) + (c+d)^2)^2$$

There is an infinite number of solutions to this equation as solving for the variables involves an elliptic curve. Small ones are,

$$a,b,c,d = -2634, 955, 1770, 5400$$

$$a,b,c,d = -31764, 7590, 27385, 48150$$

Equal Sums of two Squares

One way to generate solutions to $a^2 + b^2 = c^2 + d^2$ is to parametrize a, b, c, d in terms of integers m, n, p, q as follows:

$$(m^2 + n^2)(p^2 + q^2) = (mp - nq)^2 + (np + mq)^2 = (mp + nq)^2 + (np - mq)^2.$$

Equal Sums of Two Fourth Powers

Given two sets of Pythagorean triples,

$$(a^2 - b^2)^2 + (2ab)^2 = (a^2 + b^2)^2$$

$$(c^2 - d^2)^2 + (2cd)^2 = (c^2 + d^2)^2$$

the problem of finding equal products of a non-hypotenuse side and the hypotenuse,

$$(a^2 - b^2)(a^2 + b^2) = (c^2 - d^2)(c^2 + d^2)$$

is easily seen to be equivalent to the equation,

$$a^4 - b^4 = c^4 - d^4$$

and was first solved by Euler as $a,b,c,d = 133, 59, 158, 134$. Since he showed this is a rational point in an elliptic curve, then there is an infinite number of solutions. In fact, he also found a 7th degree polynomial parameterization.

Descartes' Circle Theorem

For the case of Descartes' circle theorem where all variables are squares,

$$2(a^4 + b^4 + c^4 + d^4) = (a^2 + b^2 + c^2 + d^2)^2$$

Euler showed this is equivalent to three simultaneous Pythagorean triples,

$$(2ab)^2 + (2cd)^2 = (a^2 + b^2 - c^2 - d^2)^2$$

$$(2ac)^2 + (2bd)^2 - (a^2 - b^2 + c^2 - d^2)^2$$

$$(2ad)^2 + (2bc)^2 = (a^2 - b^2 - c^2 + d^2)^2$$

There is also an infinite number of solutions, and for the special case when $a + b = c$, , then the equation simplifies to,

$$4(a^2 + ab + b^2) = d^2$$

with small solutions as $a,b,c,d = 3, 5, 8, 14$ and can be solved as binary quadratic forms.

Almost-isosceles Pythagorean Triples

No Pythagorean triples are isosceles, because the ratio of the hypotenuse to either other side is $\sqrt{2}$, but $\sqrt{2}$ cannot be expressed as the ratio of 2 integers.

There are, however, right-angled triangles with integral sides for which the lengths of the non-hypotenuse sides differ by one, such as,

$$3^2 + 4^2 = 5^2$$

$$20^2 + 21^2 = 29^2$$

and an infinite number of others. They can be completely parameterized as,

$$(\tfrac{x-1}{2})^2 + (\tfrac{x+1}{2})^2 = y^2$$

where $\{x, y\}$ are the solutions to the Pell equation $x^2 - 2y^2 = -1.$.

When it is the longer non-hypotenuse side and hypotenuse that differ by one, such as in

$$5^2 + 12^2 = 13^2$$

$$7^2 + 24^2 = 25^2$$

then the complete solution is

$$(2m+1)^2 + (2m^2 + 2m)^2 = (2m^2 + 2m + 1)^2$$

which also shows that all odd numbers (greater than 1) appear in a primitive Pythagorean triple.

Fibonacci Numbers in Pythagorean Triples

Starting with 5, every second Fibonacci number is the length of the hypotenuse of a right triangle with integer sides, or in other words, the largest number in a Pythagorean triple. The length of the longer leg of this triangle is equal to the sum of the three sides of the preceding triangle in this series of triangles, and the shorter leg is equal to the difference between the preceding bypassed Fibonacci number and the shorter leg of the preceding triangle.

Generalizations

There are several ways to generalize the concept of Pythagorean triples.

Pythagorean Quadruple

A set of four positive integers a, b, c and d such that $a^2 + b^2 + c^2 = d^2$ is called a Pythagorean quadruple. The simplest example is (1, 2, 2, 3), since $1^2 + 2^2 + 2^2 = 3^2$. The next simplest (primitive) example is (2, 3, 6, 7), since $2^2 + 3^2 + 6^2 = 7^2$.

All quadruples are given by the formula

$$(m^2 + n^2 - p^2 - q^2)^2 + (2mq + 2np)^2 + (2nq - 2mp)^2 = (m^2 + n^2 + p^2 + q^2)^2.$$

Pythagorean *n*-tuple

Using the simple algebraic identity,

$$(x_1^2 - x_0)^2 + (2x_1)^2 x_0 = (x_1^2 + x_0)^2$$

for arbitrary x_0, x_1, it is easy to prove that the square of the sum of n squares is itself the sum of n squares by letting $x_0 = x_2^2 + x_3^2 + \ldots + x_n^2$ and then distributing terms. One can see how Pythagorean triples and quadruples are just the particular cases $x_0 = x_2^2$ and $x_0 = x_2^2 + x_3^2$, respectively, and so on for other n, with quintuples given by

$$(a^2 - b^2 - c^2 - d^2)^2 + (2ab)^2 + (2ac)^2 + (2ad)^2 = (a^2 + b^2 + c^2 + d^2)^2.$$

Since the sum $F(k,m)$ of k consecutive squares beginning with m^2 is given by the formula,

$$F(k,m) = km(k-1+m) + \frac{k(k-1)(2k-1)}{6}$$

One may find values (k, m) so that $F(k,m)$ is a square, such as one by Hirschhorn where the number of terms is itself a square,

$$m = \tfrac{v^4 - 24v^2 - 25}{48}, k = v^2, F(m,k) = \tfrac{v^5 + 47v}{48}$$

and $v \geq 5$ is any integer not divisible by 2 or 3. For the smallest case $v = 5$, hence $k = 25$, this yields the well-known cannonball-stacking problem of Lucas,

$$0^2 + 1^2 + 2^2 + \ldots + 24^2 = 70^2$$

a fact which is connected to the Leech lattice.

In addition, if in a Pythagorean n-tuple ($n \geq 4$) all addends are consecutive except one, one can use the equation,

$$F(k,m) + p^2 = (p+1)^2$$

Since the second power of p cancels out, this is only linear and easily solved for as $p = \tfrac{F(k,m)-1}{2}$ though k, m should be chosen so that p is an integer, with a small example being $k = 5$, $m = 1$ yielding,

$$1^2 + 2^2 + 3^2 + 4^2 + 5^2 + 27^2 = 28^2$$

Thus, one way of generating Pythagorean n-tuples is by using, for various x,

$$x^2 + (x+1)^2 + \cdots + (x+q)^2 + p^2 = (p+1)^2,$$

where $q = n-2$ and where

$$p = \frac{(q+1)x^2 + q(q+1)x + \frac{q(q+1)(2q+1)}{6} - 1}{2}.$$

Fermat's Last Theorem

A generalization of the concept of Pythagorean triples is the search for triples of positive integers a, b, and c, such that $a^n + b^n = c^n$, for some n strictly greater than 2. Pierre de Fermat in 1637 claimed that no such triple exists, a claim that came to be known as Fermat's Last Theorem because it took longer than any other conjecture by Fermat to be proven or disproven. The first proof was given by Andrew Wiles in 1994.

$n - 1$ or n nth Powers Summing to an nth Power

Another generalization is searching for sequences of $n + 1$ positive integers for which the nth power of the last is the sum of the nth powers of the previous terms. The smallest sequences for known values of n are:

- $n = 3$: {3, 4, 5; 6}.
- $n = 4$: {30, 120, 272, 315; 353}
- $n = 5$: {19, 43, 46, 47, 67; 72}
- $n = 7$: {127, 258, 266, 413, 430, 439, 525; 568}
- $n = 8$: {90, 223, 478, 524, 748, 1088, 1190, 1324; 1409}

For the n=3 case, in which $x^3 + y^3 + z^3 = w^3$ called the Fermat cubic, a general formula exists giving all solutions.

A slightly different generalization allows the sum of $(k + 1)$ nth powers to equal the sum of $(n - k)$ nth powers. For example:

- (n = 3): $1^3 + 12^3 = 9^3 + 10^3$, made famous by Hardy's recollection of a conversation with Ramanujan about the number 1729 being the smallest number that can be expressed as a sum of two cubes in two distinct ways.

There can also exist $n - 1$ positive integers whose nth powers sum to an nth power (though, by Fermat's last theorem, not for $n = 3$); these are counterexamples to Euler's sum of powers conjecture. The smallest known counterexamples are

- $n = 4$: (95800, 217519, 414560; 422481)
- $n = 5$: (27, 84, 110, 133; 144)

Heronian Triangle Triples

A Heronian triangle is commonly defined as one with integer sides whose area is also an integer, and we shall consider Heronian triangles with *distinct* integer sides. The lengths of the sides of such a triangle form a Heronian triple (a, b, c) provided $a < b < c$. Clearly, any Pythagorean triple is a Heronian triple, since in a Pythagorean triple at least one of the legs a, b must be even, so that the area $ab/2$ is an integer. Not every Heronian triple is a Pythagorean triple, however, as the example (4, 13, 15) with area 24 shows.

If (a, b, c) is a Heronian triple, so is (ma, mb, mc) where m is any positive integer greater than one. The Heronian triple (a, b, c) is primitive provided a, b, c are pairwise relatively prime (as with a Pythagorean triple). Here are a few of the simplest primitive Heronian triples that are not Pythagorean triples:

(4, 13, 15) with area 24

(3, 25, 26) with area 36

(7, 15, 20) with area 42

(6, 25, 29) with area 60

(11, 13, 20) with area 66

(13, 14, 15) with area 84

(13, 20, 21) with area 126

By Heron's formula, the extra condition for a triple of positive integers (a, b, c) with $a < b < c$ to be Heronian is that

$$(a^2 + b^2 + c^2)^2 - 2(a^4 + b^4 + c^4)$$

or equivalently

$$2(a^2b^2 + a^2c^2 + b^2c^2) - (a^4 + b^4 + c^4)$$

be a nonzero perfect square divisible by 16.

Application to Cryptography

Primitive Pythagorean triples have been used in cryptography as random sequences and for the generation of keys.

Formulas for Generating Pythagorean Triples

A Simple two-unit-fraction Method of Generating PTs

This is a very simple method of generating Pythagorean triangles. It is based on forming the sum of two unit fractions for either consecutive odd numbers or consecutive even numbers.

odd A	next B–A+2	$^1/_A+^1/_B$	Hyp
1	3	4/3	5
3	5	8/15	17
5	7	12/35	37
7	9	16/63	65
...			

Two Consecutive odd Unit Fractions

Take two odd numbers that differ by two, such as 3 and 5. Make them into unit fractions: $^1/_3$ and $^1/_5$ and add them: $^1/_3 + {}^1/_5 = {}^8/_{15}$ The two numbers in the sum are always two sides of a primitive Pythagorean triangle! Here the Pythagorean triangle is 8, 15, 17.

even A	next B=A+2	$^1/_A+^1/_B$	Hyp
2	4	3/4	5
4	6	5/12	13
6	8	7/24	25
8	10	9/40	41
	...		

Two Consecutive Even unit Fractions

Take two even numbers that differ by two, such as 2 and 4. Make them into unit fractions: $^1/_2$ and 1/4 and add them: $^1/_2 + {}^1/_4 = {}^3/_4$ The two numbers in the reduced sum are always two sides of a primitive Pythagorean triangle. Here it is the 3, 4, 5 triangle.

However, neither using the two odds nor the two evens generate all the primitive Pythagorean triangles, but there is a method using two fractions that *does*.

The *Two-Fractions* method of Generating Pythagorean Triples

Here is a very simple way of generating as many Pythagorean triangle as you want:

Method:		Example 1:			Example 2:	
Take any two fractions (or whole numbers) whose *product is 2* Notice that the fractions do not have to be in their lowest form:		1/3	6		4/2	2/2
Add 2 to each fraction:		7/3	8		8/2	6/2
Cross multiply to turn both into whole numbers		7	24		16	12
These are two sides of a Pythagorean triangle:		7	24		16	12
To find the third, add the squares of these two numbers:		7^2+24^2 $= 49 + 576$ $= 625$			16^2+12^2 $=256+144$ $=400$	
and take the square-root to find the hypotenuse: to get the Pythagorean triangle:		$\sqrt{625} = 25$ 7 24 25			$\sqrt{400} = 20$ 16 12 20	

This works for any two fractions whose *product* is 2 and always generates a Pythagorean triangle:

Start with		to get:
1	2	3 4 5
2/2 1	2 4/2	6 8 10

1/2 2/4	8/2 4	5 12 13
3/3 1	2 6/3	9 12 15
2/3	3	8 15 17
4/4 2/2 1	2 4/2 8/4	12 16 20
1/3	6	7 24 25
3/2	4/3	20 21 29
1/4	8	9 40 41

In fact, all *primitive* Pythagorean triples are generated from two fractions in lowest form and all *non-primitive* Pythagorean triples are generated when at least one of the fractions is not in its lowest form.

The m,n Formula for Generating Pythagorean Triples

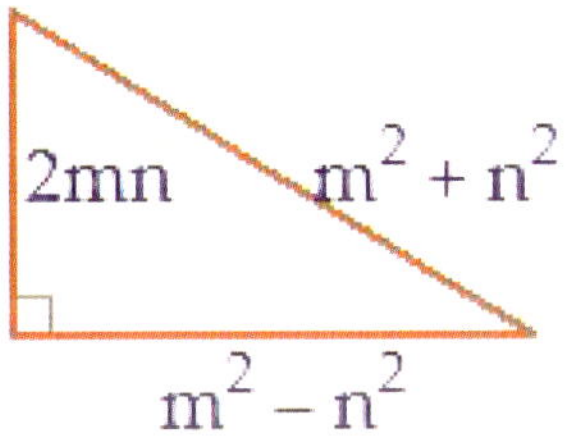

Yes - we can generate Pythagorean Triples by supplying two different positive integer values for m and n in this diagram. You can multiply out these terms and check that

$$(m^2 - n^2)^2 + (2\,m\,n)^2 = (m^2 + n^2)^2$$

Once we have found one triple, we have seen that we can generate many others by just scaling up all the sides by the same factor.

A Pythagorean triple which is not a multiple of another is calleda primitive Pythagorean triple.

So 3,4,5 and 5,12,13 are primitive Pythagorean triples but 6,8,10 and 333,444,555 and 50,120,130 are not.

Are all the Pythagorean Triples Generated by m, n?

The bad news is that the answer is "No", but the good news is that *all primitive Pythagorean triples* are generated by some m, n values in the formula above.

The formula using m, n will not give all triples since it misses some of the non-primitive ones, such as 9, 12, 15. This is a Pythagorean triple since, as a triangle, is it just three times the 3, 4, 5 triangle (by which we mean that we just multiply the lengths of each side of a 3, 4, 5 triangle by 3, which we already know is right-angled).

But 9, 12, 15 is missed by our m, n formula because:

Our formula said m and n were positive whole numbers and the Pythagorean triple was

$$m^2 - n^2,\ 2mn,\ m^2 + n^2$$

and, since we want (positive) whole number values in our triple, then m > n (otherwise the first number in the triple is negative).

The 2mn value is one of the sides and the only even side in 9, 12, 15 is 12, so 12 = 2 m n.

Hence m n = 6. But m > n, so we can only have two cases:

1. m = 6 with n = 1 OR
2. m = 3 and n = 2

The first case gives the triple 35, 12, 37 and the second case gives 5, 12, 13, neither of which is the 9, 12, 15 triple.

There *are* two values which generate 9, 12, 15 and they are $m = 2\sqrt{2}$, $n = \sqrt{2}$.

In fact there are always m,n values for *all* PTs but they are not always integers: If m,n generates a, b, h then g×a, g×b, g×h is generated by $\sqrt{g}$ m, $\sqrt{g}$ n.

$m^2 - n^2$, 2mn, $m^2 + n^2$

All the *primitive* Pythagorean triangles are each generated once if and only if) one of m, n pair is odd and the other is even. This is often described as m and n have opposite parity. Since all the sides of the triangle are positive then we also need m>n.

The two-fraction Method and the m, n Generators Method

We can now show that the two-fractions method generates all primitive Pythagorean triples by identifying n with b and m with a + b to find two starting fractions. This is equivalent to choosing a = m − n and b = n for fractions

$$\frac{a}{b} \text{ and } \frac{2b}{a}$$

to get the *primitive* triple with m, n generators.

If the triangle is non-primitive, say it is k times a primitive triangle, then the substitutions above give the primitive triangle and we need only multiply one of the fractions by k on the top *and* on the bottom to get the non-primitive one.

Fibonacci's Method

Leonardo of Pisa described this method for generating primitive triples using the sequence of consecutive odd integers 1,3,5,7,9,11,... and the fact that the sum of the first n terms of this sequence is n^2. If k is the n-th member of this sequence then $n = (k+1)/2$.

Choose any odd square number k from this sequence ($k = a^2$) and let this square be the n-th term of the sequence. Also, let b^2 be the sum of the previous $n-1$ terms, and let c^2 be the sum of all n terms. Then we have established that $a^2 + b^2 = c^2$ and we have generated the primitive triple [*a*, *b*, *c*]. This method produces an infinite number of primitive triples, but not all of them.

EXAMPLE: Choose $k = 9 = 3^2 = a^2$. This odd square number is the fifth term of the sequence, because $5 = n = (a+1)/2$. The sum of the previous 4 terms is $b^2 = 4^2$ and the sum of all $n = 5$ terms is $c^2 = 5^2$ giving us $a^2 + b^2 = c^2$ and the primitive triple [*a*, *b*, *c*] = [3, 4, 5].

Progressions of Whole and Fractional Numbers

The German monk and mathematician Michael Stifel published the following method in 1544.

Consider the progression of whole and fractional numbers: $1\frac{1}{3}, 2\frac{2}{5}, 3\frac{3}{7}, 4\frac{4}{9}, \ldots$

The properties of this progression are: (a) the whole numbers are those of the common series and have unity as their common difference; (b) the numerators of the fractions, annexed to the whole numbers, are also the natural numbers; (c) the denominators of the fractions are the odd numbers, 3, 5, 7, 9, etc.

To calculate a Pythagorean triple select any term of this progression and reduce it to an improper fraction. For example, take the term $3\frac{3}{7}$. The improper fraction is $\frac{24}{7}$. The numbers 7 and 24 are the sides, *a* and *b*, of a right triangle, and the hypotenuse is one greater than the largest side.

For example:

$$1\tfrac{1}{3} \xrightarrow{\text{yields}} [3,4,5], \quad 2\tfrac{2}{5} \xrightarrow{\text{yields}} [5,12,13], \quad 3\tfrac{3}{7} \xrightarrow{\text{yields}} [7,24,25], \quad 4\tfrac{4}{9} \xrightarrow{\text{yields}} [9,40,41], \ \ldots$$

Jacques Ozanam republished Stifel's sequence in 1694 and added the similar sequence $1\frac{7}{8}, 2\frac{11}{12}, 3\frac{15}{16}, 4\frac{19}{20}, \ldots$ with terms derived from $n + \frac{4n+3}{4n+4}$. As before, to produce a triple from this sequence, select any term and reduce it to an improper fraction. The numerator and denominator are the sides, *a* and *b*, of a right triangle. In this case, the hypotenuse of the triple(s) produced is 2 greater than the larger side.

For example:

$$1\tfrac{7}{8} \xrightarrow{\text{yields}} [15,8,17], 2\tfrac{11}{12} \xrightarrow{\text{yields}} [35,12,37], 3\tfrac{15}{16} \xrightarrow{\text{yields}} [63,16,65], 4\tfrac{19}{20} \xrightarrow{\text{yields}} [99,20,101], \ldots$$

Together, the Stifel and Ozanam sequences produce all primitive triples of the *Plato* and *Pythagoras* families respectively. The *Fermat* family must be found by other means.

Plato: $c - b = 1$, Pythagoras: $c - a = 2$, Fermat: $|a - b| = 1$

Dickson's Method

Leonard Eugene Dickson attributes to himself the following method for generating Pythagorean triples. To find integer solutions to $x^2 + y^2 = z^2$, find positive integers *r*, *s*, and *t* such that $r^2 = 2st$ is a perfect square.

Then:

$$x = r + s, y = r + t, z = r + s + t.$$

From this we see that r is any even integer and that s and t are factors of $\frac{r^2}{2}$. All Pythagorean triples may be found by this method. When s and t are coprime, the triple will be primitive. A simple proof of Dickson's method has been presented by Josef Rukavicka.

Example: Choose $r = 6$. Then $\frac{r^2}{2} = 18$. The three factor-pairs of 18 are: (1, 18), (2, 9), and (3, 6). All three factor pairs will produce triples using the above equations.

> $s = 1, t = 18$ produces the triple [7, 24, 25] because $x = 6 + 1 = 7,\ y = 6 + 18 = 24,\ z = 6 + 1 + 18 = 25$.
>
> $s = 2, t = 9$ produces the triple [8, 15, 17] because $x = 6 + 2 = 8,\ y = 6 + 9 = 15,\ z = 6 + 2 + 9 = 17$.
>
> $s = 3, t = 6$ produces the triple [9, 12, 15] because $x = 6 + 3 = 9,\ y = 6 + 6 = 12,\ z = 6 + 3 + 6 = 15$. (Since s and t are not coprime, this triple is not primitive.)

Generalized Fibonacci sequence

Method I

For Fibonacci numbers starting with $F_1 = 0$ and $F_2 = 1$ and with each succeeding Fibonacci number being the sum of the preceding two, one can generate a sequence of Pythagorean triples starting from $(a_3, b_3, c_3) = (4, 3, 5)$ via

$$(a_n, b_n, c_n) = (a_{n-1} + b_{n-1} + c_{n-1}, F_{2n-1} - b_{n-1}, F_{2n})$$

for $n \geq 4$.

Method II

A Pythagorean triple can be generated using any two positive integers by the following procedures using generalized Fibonacci sequences.

For initial positive integers h_n and h_{n+1}, if $h_n + h_{n+1} = h_{n+2}$ and $h_{n+1} + h_{n+2} = h_{n+3}$, then

$$(2h_{n+1}h_{n+2}, h_n h_{n+3}, 2h_{n+1}h_{n+2} + h_n^2)$$

is a Pythagorean triple.

Method III

The following is a matrix-based approach to generating primitive triples with generalized Fibonacci sequences. Start with a 2 × 2 array and insert two coprime positive integers (q, q') in the top row. Place the even integer (if any) in the left-hand column.

$$\begin{bmatrix} q & q' \\ \bullet & \bullet \end{bmatrix}$$

Now apply the following "Fibonacci rule" to get the entries in the bottom row:

$$\begin{matrix} q'+q=p \\ q+p=p' \end{matrix} \rightarrow \begin{bmatrix} q & q' \\ p & p' \end{bmatrix}$$

Such an array may be called a "Fibonacci Box". Note that q', q, p, p' is a generalized Fibonacci sequence. Taking column, row, and diagonal products we obtain the sides of triangle [a, b, c], its area A, and its perimeter P, as well as the radii r_i of its incircle and three excircles as follows:

$$a = 2qp$$
$$b = q'p'$$
$$c = pp' - qq' = qp' + q'p$$

$$\text{radii} \rightarrow (r_1 = qq', r_2 = qp', r_3 = q'p, r_4 = pp')$$
$$A = qq'pp'$$
$$P = r_1 + r_2 + r_3 + r_4$$

The half-angle tangents at the acute angles are q/p and q'/p'.

Example:

Using coprime integers 9 and 2.

$$\begin{bmatrix} 2 & 9 \\ \bullet & \bullet \end{bmatrix} \rightarrow \begin{bmatrix} 2 & 9 \\ 11 & 13 \end{bmatrix}$$

The column, row, and diagonal products are: (columns: 22 and 117), (rows: 18 and 143), (diagonals: 26 and 99), so

$$a = 2(22) = 44$$
$$b = 117$$
$$c = (143 - 18) = (26 + 99) = 125$$

$$\text{radii} \rightarrow (r_1 = 18, \quad r_2 = 26, \quad r_3 = 99, \quad r_4 = 143)$$
$$A = 18(143) = 2574$$
$$P = (18 + 26 + 99 + 143) = 286$$

The half-angle tangents at the acute angles are 2/11 and 9/13. Note that if the chosen integers q, q' are not coprime, the same procedure leads to a non-primitive triple.

Pythagorean Triples and Descartes' Circle Equation

This method of generating *primitive Pythagorean triples* also provides integer solutions to Descartes' Circle Equation,

$$\left(k_1 + k_2 + k_3 + k_4\right)^2 = 2\left(k_1^2 + k_2^2 + k_3^2 + k_4^2\right),$$

where integer curvatures k_i are obtained by multiplying the reciprocal of each radius by the area A. The result is k_1 = pp', k_2 = qp', k_3 = q'p, k_4 = qq'. Here, the largest circle is taken as having negative curvature with respect to the other three. The largest circle (curvature k_4) may also be replaced by a smaller circle with positive curvature ($k_0 = 4pp' - qq'$).

Example:

Using the area and four radii obtained above for primitive triple [44, 117, 125] we obtain the following integer solutions to Descartes' Equation: $k_1 = 143$, $k_2 = 99$, $k_3 = 26$, $k_4 = (-18)$, and $k_0 = 554$.

A Ternary Tree: Generating All Primitive Pythagorean Triples

Each primitive Pythagorean triple corresponds uniquely to a Fibonacci Box. Conversely, each Fibonacci Box corresponds to a unique and primitive Pythagorean triple. In this section we shall use the Fibonacci Box in place of the primitive triple it represents. An infinite ternary tree containing all primitive Pythagorean triples/Fibonacci Boxes can be constructed by the following procedure.

Consider a Fibonacci Box containing two, odd, coprime integers x and y in the right-hand column.

$$\begin{bmatrix} \bullet & x \\ \bullet & y \end{bmatrix}$$

It may be seen that these integers can also be placed as follows:

$$\begin{bmatrix} \bullet & x \\ y & \bullet \end{bmatrix}, \begin{bmatrix} x & y \\ \bullet & \bullet \end{bmatrix}, \begin{bmatrix} y & x \\ \bullet & \bullet \end{bmatrix}$$

resulting in three more valid Fibonacci boxes containing x and y. We may think of the first Box as the "parent" of the next three. For example, if $x = 1$ and $y = 3$ we have:

$$\begin{bmatrix} 1 & 1 \\ 2 & 3 \end{bmatrix} \leftarrow \text{parent}$$

$$\begin{bmatrix} 2 & 1 \\ 3 & 5 \end{bmatrix}, \begin{bmatrix} 1 & 3 \\ 4 & 5 \end{bmatrix}, \begin{bmatrix} 3 & 1 \\ 4 & 7 \end{bmatrix} \leftarrow \text{children}$$

Moreover, each "child" is itself the parent of three more children which can be obtained by the same procedure. Continuing this process at each node leads to an infinite ternary tree containing all possible Fibonacci Boxes, or equivalently, to a ternary tree containing all possible primitive triples. (The tree shown here is distinct from the classic tree described by Berggren in 1934, and has many different number-theoretic properties.) Compare: "Classic Tree".

Generating Triples using Quadratic Equations

There are several methods for defining quadratic equations for calculating each leg of a Pythagorean triple. A simple method is to modify the standard Euclid equation by adding a variable x to each m and n pair. The m, n pair is treated as a constant while the value of x is varied to produce a "family" of triples based on the selected triple. An arbitrary coefficient can be placed in front of the "x" value on either m or n, which causes the resulting equation to systematically "skip" through the triples. For example, consider the triple [20, 21, 29] which can be calculated from the Euclid equations with a value of $m = 5$ and $n = 2$. Also, arbitrarily put the coefficient of 4 in front of the "x" in the "m" term.

Let $m_1 = (4x + m)$ and let $n_1 = (x + n)$

Hence, substituting the values of m and n:

$$\text{Side } A = 2m_1n_1 \quad = 2(4x+5)(x+2) \quad = 8x^2 + 26x + 20$$
$$\text{Side } B = m_1^2 - n_1^2 \quad = (4x+5)^2 - (x+2)^2 \quad = 15x^2 + 36x + 21$$
$$\text{Side } C = m_1^2 + n_1^2 \quad = (4x+5)^2 + (x+2)^2 \quad = 17x^2 + 44x + 29$$

Note that the original triple comprises the constant term in each of the respective quadratic equations. Below is a sample output from these equations. Note that the effect of these equations is to cause the "m" value in the Euclid equations to increment in steps of 4, while the "n" value increments by 1.

x	side a	side b	side c	m	n
0	20	21	29	5	2
1	54	72	90	9	3
2	104	153	185	13	4
3	170	264	314	17	5
4	252	405	477	21	6

Pythagorean Triples by use of Matrices and Linear Transformations

Let $[a, b, c]$ be a primitive triple with a odd. Then 3 new triples $[a_1, b_1, c_1]$, $[a_2, b_2, c_2]$, $[a_3, b_3, c_3]$ may be produced from $[a, b, c]$ using matrix multiplication and Berggren's three matrices A, B, C. Triple $[a, b, c]$ is termed the *parent* of the three new triples (the *children*). Each child is itself the parent of 3 more children, and so on. If one begins with primitive triple [3, 4, 5], all primitive triples will eventually be produced by application of these matrices. The result can be graphically represented as an infinite ternary tree with $[a, b, c]$ at the root node. An equivalent result may be obtained using Berggrens's three linear transformations shown below.

$$\overset{A}{\begin{bmatrix} -1 & 2 & 2 \\ -2 & 1 & 2 \\ -2 & 2 & 3 \end{bmatrix}}\begin{bmatrix} a \\ b \\ c \end{bmatrix} = \begin{bmatrix} a_1 \\ b_1 \\ c_1 \end{bmatrix}, \quad \overset{B}{\begin{bmatrix} 1 & 2 & 2 \\ 2 & 1 & 2 \\ 2 & 2 & 3 \end{bmatrix}}\begin{bmatrix} a \\ b \\ c \end{bmatrix} = \begin{bmatrix} a_2 \\ b_2 \\ c_2 \end{bmatrix}, \quad \overset{C}{\begin{bmatrix} 1 & -2 & 2 \\ 2 & -1 & 2 \\ 2 & -2 & 3 \end{bmatrix}}\begin{bmatrix} a \\ b \\ c \end{bmatrix} = \begin{bmatrix} a_3 \\ b_3 \\ c_3 \end{bmatrix}$$

Berggren's three linear transformations are:

$$\begin{array}{llll} -a+2b+2c=a_1 & -2a+b+2c=b_1 & -2a+2b+3c=c_1 & \rightarrow[a_1, b_1, c_1] \\ +a+2b+2c=a_2 & +2a+b+2c=b_2 & +2a+2b+3c=c_2 & \rightarrow[a_2, b_2, c_2] \\ +a-2b+2c=a_3 & +2a-b+2c=b_3 & +2a-2b+3c=c_3 & \rightarrow[a_3, b_3, c_3] \end{array}$$

Alternatively, one may also use 3 different matrices found by Price. These matrices *A'*, *B'*, *C'* and their corresponding linear transformations are shown below.

$$\overset{A'}{\begin{bmatrix} 2 & 1 & -1 \\ -2 & 2 & 2 \\ -2 & 1 & 3 \end{bmatrix}} \begin{bmatrix} a \\ b \\ c \end{bmatrix} = \begin{bmatrix} a_1 \\ b_1 \\ c_1 \end{bmatrix}, \quad \overset{B'}{\begin{bmatrix} 2 & 1 & 1 \\ 2 & -2 & 2 \\ 2 & -1 & 3 \end{bmatrix}} \begin{bmatrix} a \\ b \\ c \end{bmatrix} = \begin{bmatrix} a_2 \\ b_2 \\ c_2 \end{bmatrix}, \quad \overset{C'}{\begin{bmatrix} 2 & -1 & 1 \\ 2 & 2 & 2 \\ 2 & 1 & 3 \end{bmatrix}} \begin{bmatrix} a \\ b \\ c \end{bmatrix} = \begin{bmatrix} a_3 \\ b_3 \\ c_3 \end{bmatrix}$$

Price's three linear transformations are

$$\begin{array}{llll} +2a+b-c=a_1 & -2a+2b+2c=b_1 & -2a+b+3c=c_1 & \rightarrow[a_1, b_1, c_1] \\ +2a+b+c=a_2 & +2a-2b+2c=b_2 & +2a-b+3c=c_2 & \rightarrow[a_2, b_2, c_2] \\ +2a-b+c=a_3 & +2a+2b+2c=b_3 & +2a+b+3c=c_3 & \rightarrow[a_3, b_3, c_3] \end{array}$$

The 3 children produced by each of the two sets of matrices are not the same, but each set separately produces all primitive triples.

For example, using [5, 12, 13] as the parent, we get two sets of three children:

$$\begin{array}{ccc} & [5,12,13] & \\ A & B & C \\ [45,28,53] & [55,48,73] & [7,24,25] \end{array} \qquad \begin{array}{ccc} & [5,12,13] & \\ A' & B' & C' \\ [9,40,41] & [35,12,37] & [11,60,61] \end{array}$$

Area proportional to Sums of Squares

All primitive triples with $b+1=c$ and with a odd can be generated as follows:

Pythagorean triple	Semi-perimeter	Area	Incircle radius	Circumcircle radius
$(3,4,5)$	$1+2+3$	$6\times(1^2)$	1	$\frac{5}{2}$
$(5,12,13)$	$1+2+3+4+5$	$6\times(1^2+2^2)$	2	$\frac{13}{2}$
$(7,24,25)$	$1+2+3+4+5+6+7$	$6\times(1^2+2^2+3^2)$	3	$\frac{25}{2}$
....	....	....	....	....
$\left(a,\frac{a^2-1}{2},\frac{a^2+1}{2}\right)$	$1+2+\ldots+a$	$6\times\left[1^2+2^2+\cdots+\left(\frac{a-1}{2}\right)^2\right]$	$\left(\frac{a-1}{2}\right)$	$\frac{c}{2}$

Height-excess Enumeration Theorem

Wade and Wade first introduced the categorization of Pythagorean triples by their height, defined as c - b, linking 3,4,5 to 5,12,13 and 7,24,25 and so on.

McCullough and Wade extended this approach, which produces all Pythagorean triples when $k > \frac{h\sqrt{2}}{d}$ Write a positive integer h as pq^2 with p square-free and q positive. Set $d = 2pq$ if p is odd, or $d = pq$ if p is even. For all pairs (h,k) of positive integers, the triples are given by

$$(h+dk, dk+\frac{(dk)^2}{2h}, h+dk+\frac{(dk)^2}{2h}).$$

The primitive triples occur when $\gcd(k, h) = 1$ and either $h=q^2$ with q odd or $h=2q^2$.

Pythagorean Trigonometric Identity

The most common trigonometric identities are those involving the Pythagorean Theorem.

When studying the unit circle (radius of 1), it was observed that a point on the unit circle (a vertex of the right triangle) can be represented by the coordinates $(\cos\theta, \sin\theta)$. Since the legs of the right triangle in the unit circle have the values of $\sin\theta$ and $\cos\theta$, the Pythagorean Theorem can be used to obtain $\sin^2\theta + \cos^2\theta = 1$.

This well-known equation is called a Pythagorean Identity. It is true for all values of θ in the unit circle.

$$\sin^2\theta + \cos^2\theta = 1$$

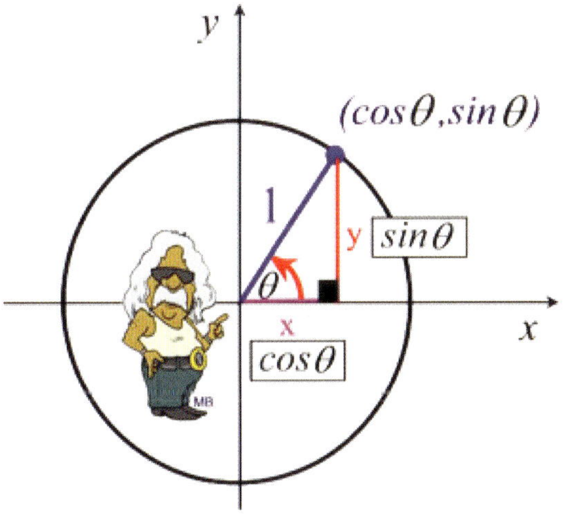

This well-known equation is called a Pythagorean Identity. It is true for all values of θ in the unit circle.

Using this first Pythagorean Identity, two additional Pythagorean Identities can be created.

$\sin^2\theta+\cos^2\theta=1$	• Start with this first Pythagorean Identity. • Divide each term by $\cos^2\theta$.
$\dfrac{\sin^2\theta}{\cos^2\theta}+\dfrac{\cos^2\theta}{\cos^2\theta}=\dfrac{1}{\cos^2\theta}$	• We know $\dfrac{\sin\theta}{\cos\theta}=\tan\theta$ and $\dfrac{1}{\cos\theta}=\sec\theta$
$\tan^2\theta+1=\sec^2\theta$	• Substitute and simplify.

We now have a second Pythagorean Identity:

$$\tan^2\theta+1=\sec^2\theta$$

It should be noted that there are values of θ for which tangent and secant are undefined.

Remember that $\tan\theta=\dfrac{\sin\theta}{\cos\theta}$ and when $\cos\theta=0, \tan\theta$ is undefined

Also, $\sec\theta=\dfrac{1}{\cos\theta}$ and the same undefined situation may exist.

If we divide by a different value, we can arrive at the third identity:

$\sin^2\theta+\cos^2\theta=1$	• Start with this first Pythagorean Identity. • Divide each term by $\sin^2\theta$.
$\dfrac{\sin^2\theta}{\sin^2\theta}+\dfrac{\cos^2\theta}{\sin^2\theta}=\dfrac{1}{\sin^2\theta}$	• We know $\dfrac{\cos\theta}{\sin\theta}=\cot\theta$ and $\dfrac{1}{\sin\theta}=\csc\theta$.
$1+\cot^2\theta=\csc^2\theta$	• Substitute and simplify.

The third Pythagorean Identity is:

$$1+\cot^2\theta=\csc^2\theta$$

Remember $\cot\theta=\dfrac{\cos\theta}{\sin\theta}$ and $\csc\theta=\dfrac{1}{\sin\theta}$ are undefined when $\sin\theta=0$.

Example

Use Pythagorean Identities to find missing trigonometric values.

if $\sec u=-\dfrac{3}{2}$ and $\tan u>0$,

find the value of the other trig functions

Solution:

The easiest Pythagorean Identity to work with is

$\sin^2\theta+\cos^2\theta=1$.

We know $\cos u = \frac{1}{\sec u} = \frac{1}{-3/2} = -\frac{2}{3}$.

Now, $\sin^2\theta + \cos^2\theta = 1$, gives $\sin^2 u + \left(-\frac{2}{3}\right)^2 = 1$

$$\sin^2 u = 1 - \frac{4}{9} = \frac{5}{9}$$

$$\sin u = \pm\frac{\sqrt{5}}{3}$$

since $\cos u < 0$ and $\tan u > 0$, u is in Quadrant III which means $\sin u = -\frac{\sqrt{5}}{3}$.

Now, $\sin u = \frac{-\sqrt{5}}{3}$ And, $\csc u = \frac{3}{-\sqrt{5}}$

$\cos u = \frac{-2}{3}$ $\sec u = \frac{3}{-2}$

$\tan u = \frac{-\sqrt{5}}{-2} = \frac{\sqrt{5}}{2}$ $\cot u = \frac{2}{\sqrt{5}}$

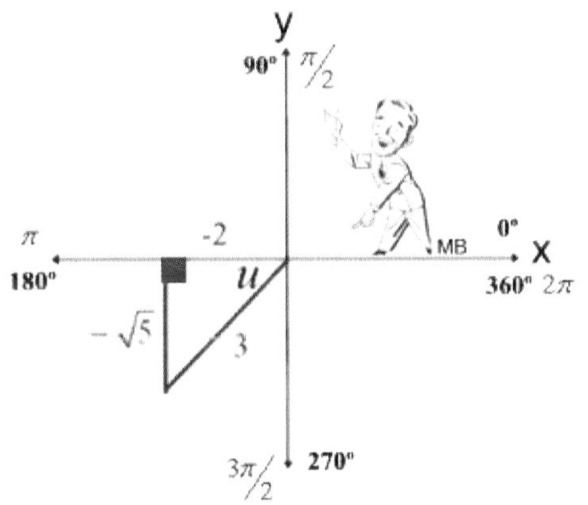

Example:

Use Pythagorean Identities to help simplify trig expressions

Simplify: $\sin x \cos^2 x - \sin x$

Solution: Since this expression contains sine and cosine, utilize $\sin^2\theta + \cos^2\theta = 1$ or its variations.

Start by factoring:

$$\sin x\cos^2 x - \sin x = \sin x\left(\cos^2 x - 1\right)$$

$$\sin x\left(-\sin^2 x\right)$$

$$-\sin^3 x$$

Use this substitution:

$$\sin^2 x + \cos^2 x = 1$$

gives

$$\cos^2 x - 1 = -\sin^2 x$$

Example: Use Pythagorean Identities to help simplify a trig expression into a factorable form

Express $csc^2 x - cotx - 3$ in factored form.

Solution:

Start by substituting:

$$\begin{aligned}\csc^2 x - \cot x - 3 &= 1 + \cot^2 x - \cot x - 3 \\ &\cot^2 x - \cot x - 2 \\ &(\cot x - 2)(\cot x + 1)\end{aligned}$$

Use this Pythagorean Identity:

$$1 + \cot^2 x = \csc^2 x$$

Ptolemy's Theorem

Ptolemy's Theorem is a relation in Euclidean geometry between the four sides and two diagonals of a cyclic quadrilateral (i.e., a quadrilateral whose vertices lie on a common circle). The theorem is named after the Greek astronomer and mathematician Ptolemy (Claudius Ptolemaeus).

If the quadrilateral is given with its four vertices A, B, C, and D in order, then the theorem states that:

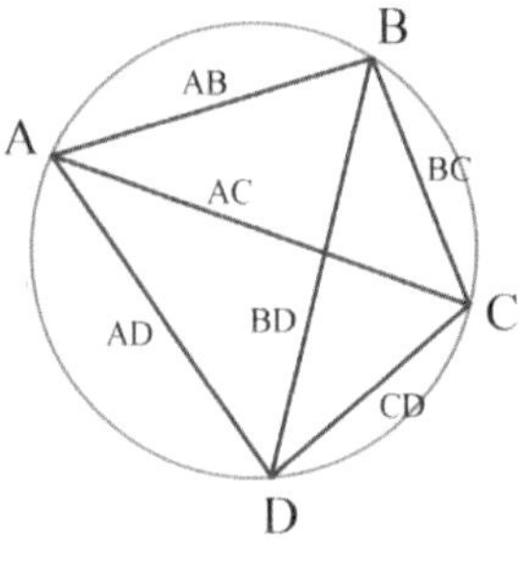

$$|AC| \cdot |BD| = |AB| \cdot |CD| + |BC| \cdot |AD|$$

This relation may be verbally expressed as follows:

If a quadrilateral is inscribed in a circle then the sum of the products of its two pairs of opposite sides is equal to the product of its diagonals.

Proof

Geometric proof of Ptolemy's Theorem

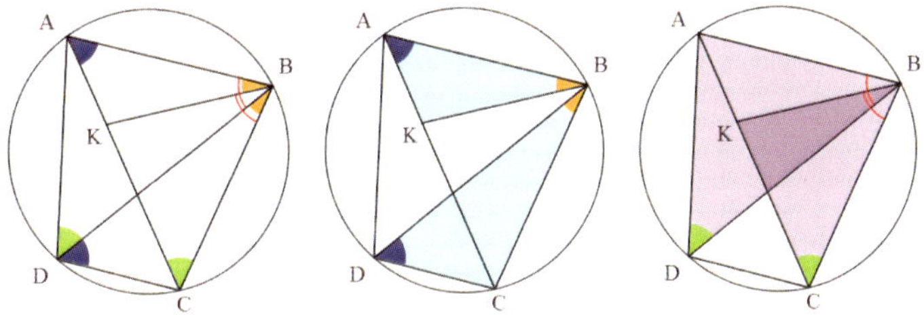

1. Let ABCD be a cyclic quadrilateral.

2. Note that on the chord BC, the inscribed angles ∠BAC = ∠BDC, and on AB, ∠ADB = ∠ACB.

3. Construct K on AC such that ∠ABK = ∠CBD;

[Note that: ∠ABK + ∠CBK = ∠ABC = ∠CBD + ∠ABD ⇒ ∠CBK = ∠ABD.]

4. Now, by common angles △ABK is similar to △DBC, and likewise △KBC ~ △ABD.

5. Thus, $\frac{|AK|}{|AB|} = \frac{|DC|}{|DB|}$ and $\frac{|KC|}{|BC|} = \frac{|AD|}{|BD|}$ due to the similarities noted above:

$[\Delta ABK \sim \Delta DBC \text{ and } \Delta KBC \sim \Delta ABD]$

1. *So* $|AK| \cdot |DB| = |AB| \cdot |DC|$, *and* $|KC| \cdot |BD| = |BC| \cdot |AD|$;
2. *Adding,* $|AK| \cdot |DB| + |KC| \cdot |BD| = |AB| \cdot |DC| + |BC| \cdot |AD|$;
3. *Equivalently,* $(|AK| + |KC|) \cdot |BD| = |AB| \cdot |CD| + |BC| \cdot |AD|$;
4. *But* $|AK| + |KC| = |AC|$, *so*
5. $|AC| \cdot |BD| = |AB| \cdot |CD| + |BC| \cdot |DA|$; *Q.E.D.*

Proof by Trigonometric Identities

Let the inscribed angles subtended by AB, BC and CD be, respectively, α, β and γ, and the radius of the circle be R, then we have $AB = 2R\sin\alpha$, $BC = 2R\sin\beta$, $CD = 2R\sin\gamma$, $AD = 2R\sin(180 - (\alpha+\beta+\gamma))$, $AC = 2R\sin(\alpha+\beta)$ and $BD = 2R\sin(\beta+\gamma)$, and the original equality to be proved is transformed to

$$\sin(\alpha+\beta)\sin(\beta+\gamma) = \sin\alpha\sin\gamma + \sin\beta\sin(\alpha+\beta+\gamma)$$

from which the factor $4R^2$ has disappeared by dividing both sides of the equation by it.

Now by using the sum formulae, $\sin(x+y) = \sin x\cos y + \cos x\sin y$ and $\cos(x+y) = \cos x\cos y - \sin x\sin y$,, it is trivial to show that both sides of the above equation are equal to

$$\sin\alpha\sin\beta\cos\beta\cos\gamma + \sin\alpha\cos^2\beta\sin\gamma$$
$$+\cos\alpha\sin^2\beta\cos\gamma + \cos\alpha\sin\beta\cos\beta\sin\gamma.$$

Proof by Inversion

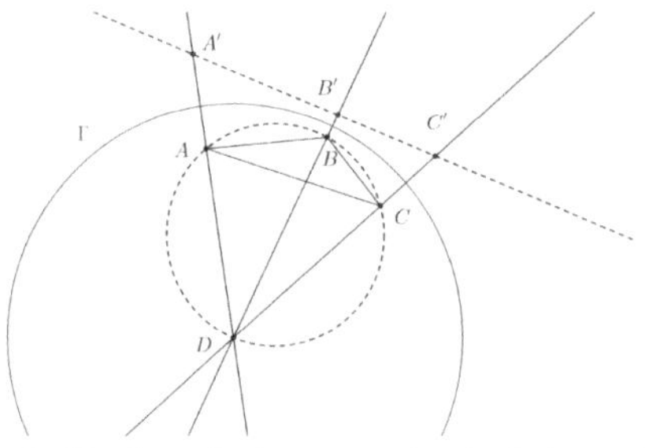

Proof of Ptolemy's theorem via circle inversion

Choose an auxiliary circle Γ centered at D with respect to which the circumcircle of ABCD is inverted into a line (see figure). Then $A'B' + B'C' = A'C'$ Without loss of generality Γ has radius 1. Then $A'B', B'C'$ and $A'C'$ can be expressed as $\frac{AB}{DA \cdot DB}, \frac{BC}{DB \cdot DC}, \frac{AC}{DA \cdot DC}$ respectively. Multiplying previous relation by $DA \cdot DB \cdot DC$ yields Ptolemy's equality.

Corollaries

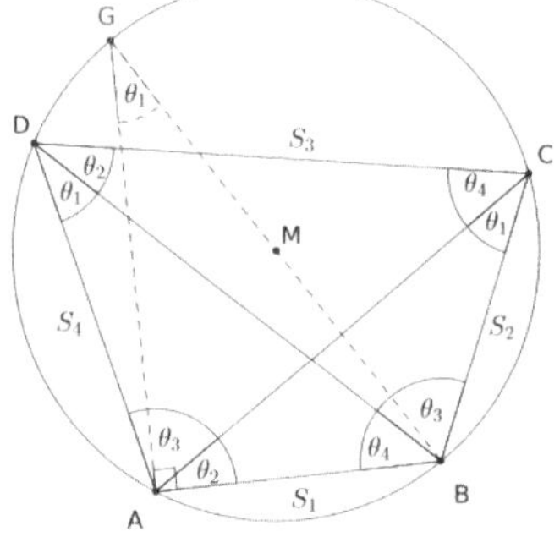

$|S_1| = \sin(\theta_1)$

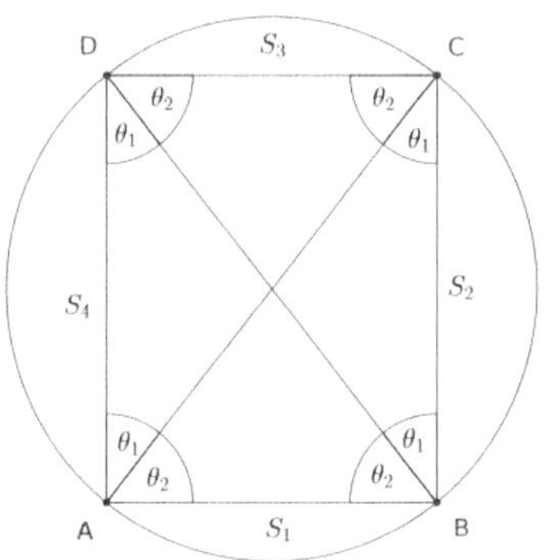

Corollary: Pythagoras's theorem

In the case of a circle of unit diameter the sides S_1, S_2, S_3, S_4 of any cyclic quadrilateral ABCD are numerically equal to the sines of the angles $\theta_1, \theta_2, \theta_3$ and θ_4 which they subtend. Similarly the diagonals are equal to the sine of the sum of whichever pair of angles they subtend. We may then write Ptolemy's Theorem in the following trigonometric form:

$$\sin\theta_1 \sin\theta_3 + \sin\theta_2 \sin\theta_4 = \sin(\theta_3 + \theta_2)\sin(\theta_1 + \theta_2)$$

Applying certain conditions to the subtended angles $\theta_1, \theta_2, \theta_3$ and θ_4 it is possible to derive a number of important corollaries using the above as our starting point. In what follows it is important to bear in mind that the sum of angles $\theta_1 + \theta_2 + \theta_3 + \theta_4 = 180^\circ$.

Corollary 1. Pythagoras's Theorem

Let $\theta_1 = \theta_3$ and $\theta_2 = \theta_4$. Then $\theta_1 + \theta_2 = \theta_3 + \theta_4 = 90^\circ$ (since opposite angles of a cyclic quadrilateral are supplementary). Then:

$$\sin\theta_1 \sin\theta_3 + \sin\theta_2 \sin\theta_4 = \sin(\theta_3 + \theta_2)\sin(\theta_3 + \theta_4)$$

$$\sin^2\theta_1 + \sin^2\theta_2 = \sin^2(\theta_1 + \theta_2)$$

$$\sin^2\theta_1 + \cos^2\theta_1 = 1$$

Corollary 2. The Law of Cosines

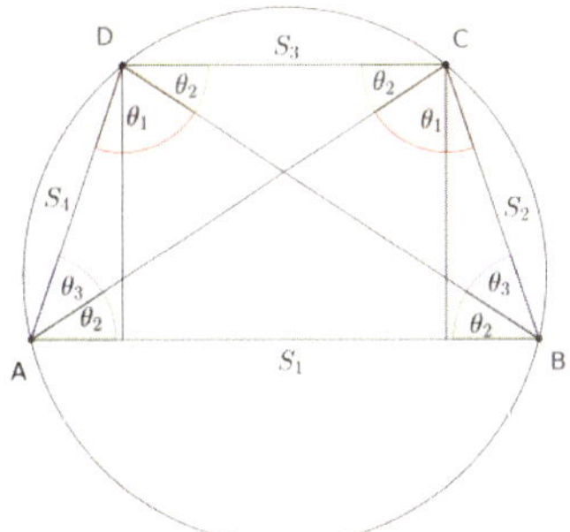

Corollary 2: the law of cosines

Let $\theta_2 = \theta_4$. The rectangle of corollary 1 is now a symmetrical trapezium with equal diagonals and a pair of equal sides. The parallel sides differ in length by $2x$ units where:

$$x = S_2 \cos(\theta_2 + \theta_3)$$

It will be easier in this case to revert to the standard statement of Ptolemy's theorem:

$$S_1S_3 + S_2S_4 = \overline{AC} \cdot \overline{BD}$$

$$\Rightarrow S_1S_3 + S_2^2 = \overline{AC}^2$$

$$\Rightarrow S_1[S_1 - 2S_2\cos(\theta_2 + \theta_3)] + S_2^2 = \overline{AC}^2$$

$$\Rightarrow S_1^2 + S_2^2 - 2S_1S_2\cos(\theta_2 + \theta_3) = \overline{AC}^2$$

The cosine rule for triangle ABC.

Corollary 3. Compound Angle sine (+)

Let

$$\theta_1 + \theta_2 = \theta_3 + \theta_4 = 90^\circ.$$

Then

$$\sin\theta_1 \sin\theta_3 + \sin\theta_2 \sin\theta_4 = \sin(\theta_3 + \theta_2)\sin(\theta_3 + \theta_4)$$

Therefore,

$$\cos\theta_2 \sin\theta_3 + \sin\theta_2 \cos\theta_3 = \sin(\theta_3 + \theta_2)\times 1$$

Formula for compound angle sine (+).

Corollary 4. Compound Angle Sine (−)

Let $\theta_1 = 90^\circ$. Then $\theta_2 + (\theta_3 + \theta_4) = 90^\circ$. Hence,

$$\sin\theta_1 \sin\theta_3 + \sin\theta_2 \sin\theta_4 = \sin(\theta_3 + \theta_2)\sin(\theta_3 + \theta_4)$$

$$\sin\theta_3 + \sin\theta_2 \cos(\theta_2 + \theta_3) = \sin(\theta_3 + \theta_2)\cos\theta_2$$

$$\sin\theta_3 = \sin(\theta_3 + \theta_2)\cos\theta_2 - \cos(\theta_2 + \theta_3)\sin\theta_2$$

Formula for compound angle sine (−).

This derivation corresponds to the Third Theorem as chronicled by Copernicus following Ptolemy in Almagest. In particular if the sides of a pentagon (subtending 36° at the circumference) and of a hexagon (subtending 30° at the circumference) are given, a chord subtending 6° may be calculated. This was a critical step in the ancient method of calculating tables of chords.

Corollary 5. Compound Angle Cosine (+)

This corollary is the core of the Fifth Theorem as chronicled by Copernicus following Ptolemy in Almagest.

Let $\theta_3 = 90^\circ$. Then $\theta_1 + (\theta_2 + \theta_4) = 90^\circ$. Hence

$$\sin\theta_1 \sin\theta_3 + \sin\theta_2 \sin\theta_4 = \sin(\theta_3 + \theta_2)\sin(\theta_3 + \theta_4)$$

$$\cos(\theta_2 + \theta_4) + \sin\theta_2 \sin\theta_4 = \cos\theta_2 \cos\theta_4$$

$$\cos(\theta_2 + \theta_4) = \cos\theta_2 \cos\theta_4 - \sin\theta_2 \sin\theta_4$$

Formula for compound angle cosine (+)

Despite lacking the dexterity of our modern trigonometric notation, it should be clear from the above corollaries that in Ptolemy's theorem (or more simply the Second Theorem) the ancient world had at its disposal an extremely flexible and powerful trigonometric tool which enabled the cognoscenti of those times to draw up accurate tables of chords (corresponding to tables of sines) and to use these in their attempts to understand and map the cosmos as they saw it. Since tables of chords were drawn up by Hipparchus three centuries before Ptolemy, we must assume he knew of the 'Second Theorem' and its derivatives. Following the trail of ancient astronomers, history records the star catalogue of Timocharis of Alexandria. If, as seems likely, the compilation of such catalogues required an understanding of the 'Second Theorem' then the true origins of the latter disappear thereafter into the mists of antiquity but it cannot be unreasonable to presume that the astronomers, architects and construction engineers of ancient Egypt may have had some knowledge of it.

Examples

Equilateral Triangle

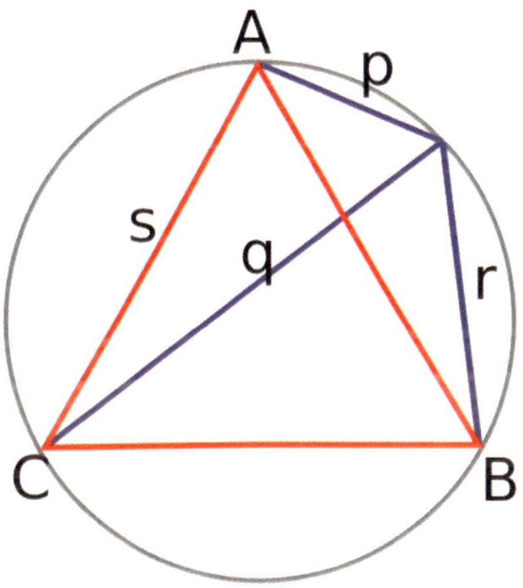

Equilateral triangle

Ptolemy's Theorem yields as a corollary a pretty theorem regarding an equilateral triangle inscribed in a circle.

Given An equilateral triangle inscribed on a circle and a point on the circle.

The distance from the point to the most distant vertex of the triangle is the sum of the distances from the point to the two nearer vertices.

Proof: Follows immediately from Ptolemy's theorem:

$$qs = ps + rs \Rightarrow q = p + r.$$

Square

Any square can be inscribed in a circle whose center is the center of the square. If the common length of its four sides is equal to a then the length of the diagonal is equal to $a\sqrt{2}$ according to the Pythagorean theorem and the relation obviously holds.

Rectangle

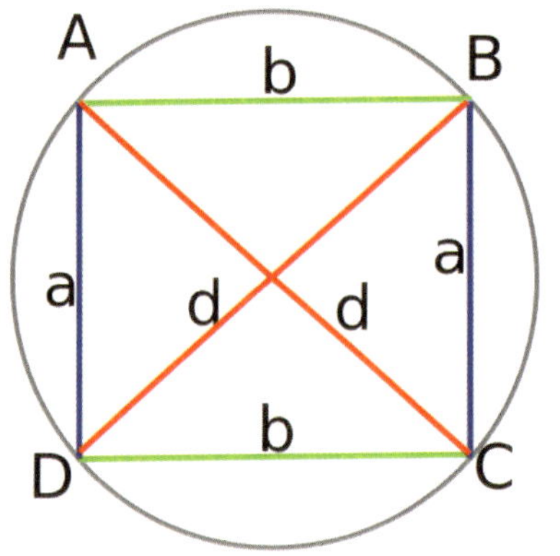

Pythagoras's theorem: *"manifestum est"*: Copernicus

More generally, if the quadrilateral is a rectangle with sides a and b and diagonal d then Ptolemy's theorem reduces to the Pythagorean theorem. In this case the center of the circle coincides with the point of intersection of the diagonals. The product of the diagonals is then d^2, the right hand side of Ptolemy's relation is the sum $a^2 + b^2$.

Copernicus – who used Ptolemy's theorem extensively in his trigonometrical work – refers to this result as a 'Porism' or self-evident corollary:

> *Furthermore it is clear (manifestum est) that when the chord subtending an arc has been given, that chord too can be found which subtends the rest of the semicircle.*

Pentagon

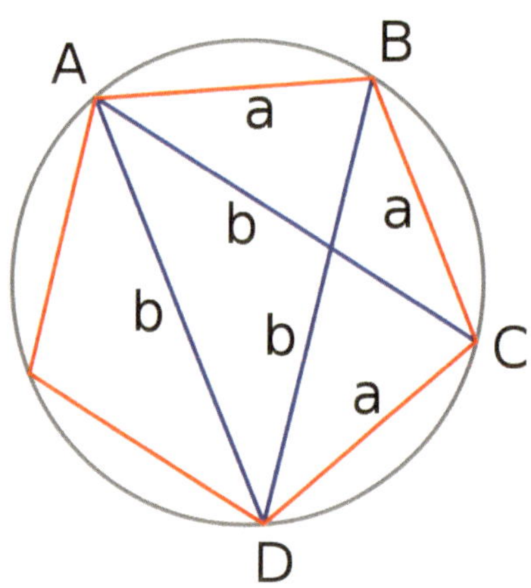

The golden ratio follows from this application of Ptolemy's theorem

A more interesting example is the relation between the length a of the side and the (common) length b of the 5 chords in a regular pentagon. In this case the relation reads $b^2 = a^2 + ab$ which yields the golden ratio

$$\varphi = \frac{b}{a} = \frac{1+\sqrt{5}}{2}.$$

Side of Decagon

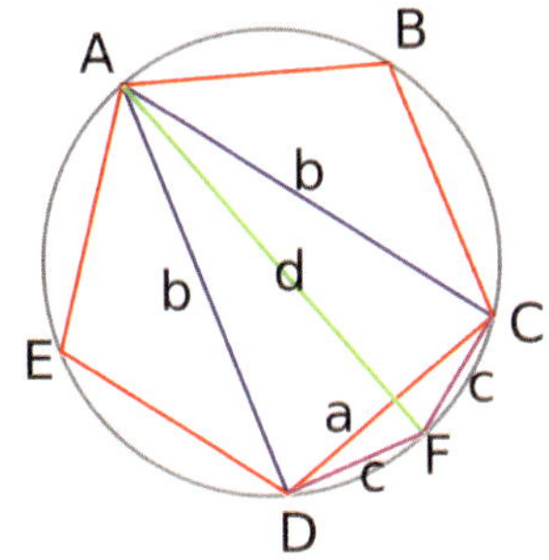

Side of the inscribed decagon

If now diameter AF is drawn bisecting DC so that DF and CF are sides c of an inscribed decagon, Ptolemy's Theorem can again be applied – this time to cyclic quadrilateral ADFC with diameter *d* as one of its diagonals:

$$ad = 2bc$$

$$\Rightarrow ad = 2\varphi ac \text{ where } \varphi \text{ is the golden ratio.}$$

$$\Rightarrow c = \frac{d}{2\varphi}.$$

whence the side of the inscribed decagon is obtained in terms of the circle diameter. Pythagoras's theorem applied to right triangle AFD then yields "b" in terms of the diameter and "a" the side of the pentagon is thereafter calculated as

$$a = \frac{b}{\varphi} = b(\varphi - 1).$$

As Copernicus (following Ptolemy) wrote,

> *"The diameter of a circle being given, the sides of the triangle, tetragon, pentagon, hexagon and decagon, which the same circle circumscribes, are also given."*

Ptolemy's Inequality

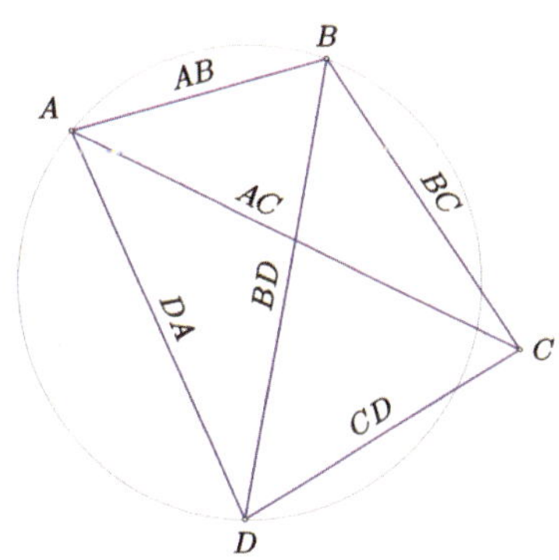

This is *not* a cyclic quadrilateral. The equality never holds here, and is unequal in the direction indicated by Ptolemy's inequality.

The equation in Ptolemy's theorem is never true with non-cyclic quadrilaterals. Ptolemy's inequality is an extension of this fact, and it is a more general form of Ptolemy's theorem. It states that, given a quadrilateral *ABCD*, then

$$\overline{AB}\cdot\overline{CD}+\overline{BC}\cdot\overline{DA}\geq\overline{AC}\cdot\overline{BD}$$

where equality holds if and only if the quadrilateral is cyclic. This special case is equivalent to Ptolemy's theorem.

Thales's Theorem

Thales' Theorem states the following:

Using the diameter of a circle as the base of a triangle with the apex on that circle, means that this triangle will be a right triangle and the diameter will be the hypotenuse of that triangle.

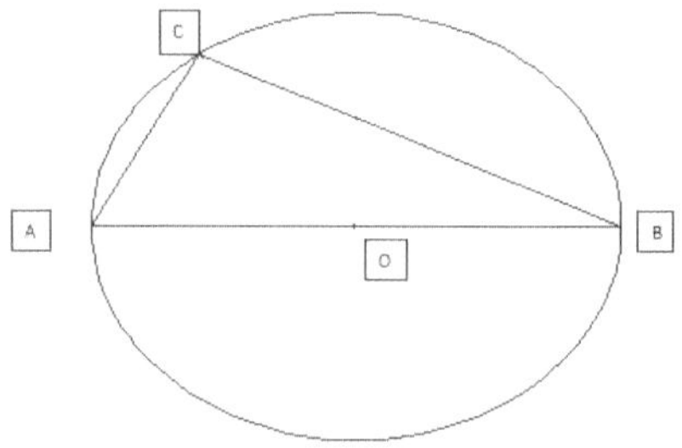

"Triangle in a Circle"

"The diameter of a circle subtends a right angle to any point on the circle". (The word "subtends" means that it "creates an angle at a distant point").

The diagram shows it more clearly. Position 'O' is the center of the circle. Line AB is the diameter of the circle. Point 'c' is on the circumference.

Thales' Theorem says that the angle ACB of the triangle ACB is a right angle regardless of where 'C' may be placed on the circumference.

To Prove Thales' Theorem

To prove Thales' Theorem, let's start by adding another line, CO, to the diagram.

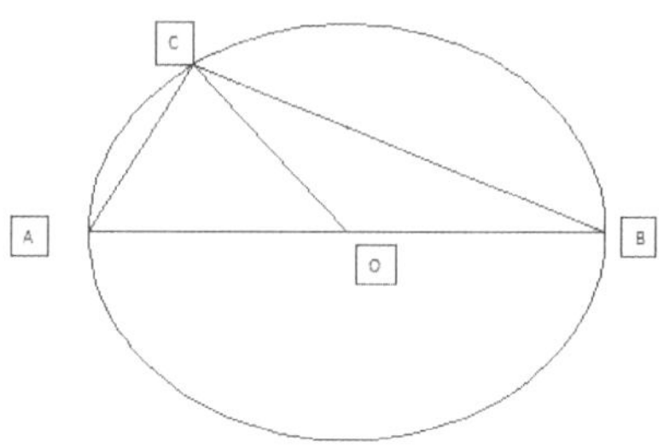

Thales Triangle with Notes

Preparing for the Proof of Thales' Theorem

The point 'O' is the center of a circle with radius of length 'r'. AB is a diameter with 'O' at the center, so length(AO) = length(OB) = r. Point C is the third point on the circumference. We have triangles OCA and OCB, and length(OC) = r also.

Triangle OCA is isosceles since length(AO) = length(CO) = r. Therefore angle (OAC) = angle(OCA); let's call it 'α' ("alpha"). Likewise, triangle OCB is isosceles since length (BO) = length (CO) = r. Therefore angle(OBC) = angle(OCB); let's call it 'β' ("beta").

Since AOB is a straight line, angle (AOB) = 180. So angle(AOC) + angle(COB) = 180. Let's record this fact as angle(COB) = 180 – angle(AOC).

The sum of the interior angles of a triangle is 180. So in triangle OCA, the sum of angles OAC + OCA + AOC = 180. As we said before, this is α+α + angle(AOC) = 180, or 2*α + angle(AOC) = 180.

Likewise, in triangle OCB, angle(OCB) + angle(OBC) + angle(COB) = 180. As noted, we can say β+β + angle(COB) = 180, or 2*β + angle(COB) = 180.

First Proof of Thales' Theorem

We also see that triangle ACB is a triangle, so its interior angles add up to 180 also. So angle (BAC) + angle (ABC) + angle (ACB) = 180. That was α+β + angle (ACB) = 180.

But clearly angle (ACB) is the sum of angle (ACO) + angle (OCB), which are values α and β. So α+β + α+β = 180. So 2*(α+β) = 180. Dividing both sides by 2 shows that α+β = 90. This is precisely what Thales' Theorem says, so it is proven.

Second Proof of Thales' Theorem

The first proof used the fact that angle (OAC) is precisely angle (BAC). The viewpoint changed from the two smaller triangles to the one triangle made up of the two smaller ones. This proof will take a different viewpoint.

First, let's remember that angle (COB) = 180 – angle (AOC). So the previous sum of triangle OBC's angles could be written as 2*β + (180 – angle (AOC)) = 180.

We have two equations: 2*α + angle (AOC) = 180, and also 2*β + (180 – angle (AOC)) = 180.

For this alternate proof, let's add the angles of the two smaller triangles. Therefore (2*α + angle(AOC)) + (2*β + (180 – angle(AOC))) – 360.

Rearrange the terms and we find (2*α + 2*β) + angle (AOC) – angle (AOC) + 180 = 360.

The beautiful part of this equation is that it explicitly shows that angle AOC does not matter: it is added and subtracted, thus cancelling itself out. The first proof simply stopped talking about angle AOC.

Bring out the common factor of "2", cancel the "angle(AOC)" terms, and subtract 180 from both sides to show 2*(α + β) = 180. Divide both sides by 2 and we have returned to Thales' Theorem, because α + β = 90.

One Practical Application of Thales' Theorem

A simple but practical application of Thales' Theorem is to find the center of a circle, assuming you can draw a couple of right-angle triangles over it.

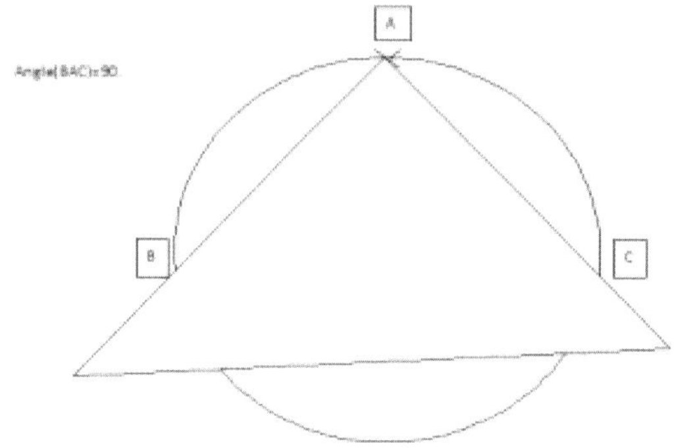

"Starting to Seek the Centre of a Circle" by Mike DeHaan

Draw a circle. Choose 1 point, "A", on the circumference. Draw a right-angle triangle with that point at the right angle, making the triangle overlap the circle and extend beyond it.

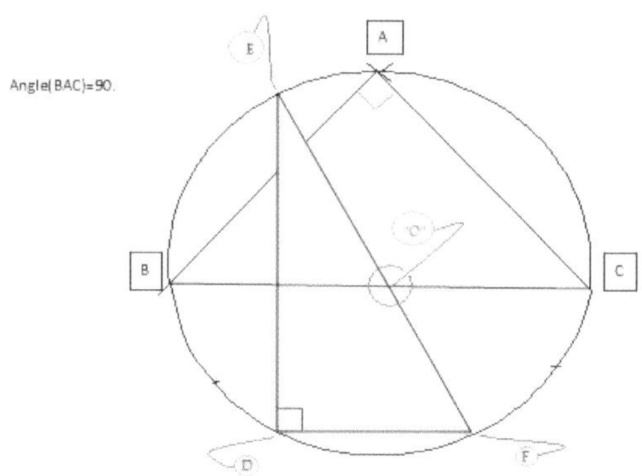

"Having Found the Centre of a Circle" by Mike DeHaan

Mark the places where the smaller sides (not the hypotenuse) cross the circle. These are points "B" and "C". The line BC is a diameter, by Thales' Theorem.

Choose a different point, "D" on the circumference, and draw a new right-angle triangle DEF. Once again, line EF will be a diameter. Their intersection is point "O", the center of the circle.

References

- Berggren, J. Lennart (2007). "Mathematics in Medieval Islam". The Mathematics of Egypt, Mesopotamia, China, India, and Islam: A Sourcebook. Princeton University Press. p. 518. ISBN 978-0-691-11485-9
- Spherical Law of Cosines: math.ucdenver.edu, Retrieved 10 June 2018
- Ozanam, Jacques (1814). "Recreations in Mathematics and Natural Philosophy". 1. G. Kearsley: 49. Retrieved 2009-11-19Law-of-tangents, trigonometry: tutorvista.com, Retrieved 25 May 2018
- J. L. Poet and D. L. Vestal, Jr. (2005). "Curious Consequences of a Miscopied Quadratic, " College Mathematics Journal 36 , 273–277Pythagoras-theorem, maths: byjus.com, Retrieved 14 April 2018

- Computing : a historical and technical perspective. Igarashi, Yoshihide,. Boca Raton, Florida. p. 78. ISBN 9781482227413. OCLC 882245835Triples-and-quadruples: plus.maths.org, Retrieved 21 May 2018
- Mr. T's Math Videos (2015-06-10), Area of a Triangle and Radius of its Circumscribed Circle, retrieved 2018-09-18
- McCullough, Darryl, and Wade, Elizabeth, "Recursive enumeration of Pythagorean triples", College Mathematics Journal 34, March 2003, 107-111

5

Trigonometric Identities

Trigonometric identities are based on trigonometric functions. They are equalities that hold for every value of a variable. The topics elaborated in this chapter will help in gaining a better perspective about the different trigonometric identities such as tangent half-angle formula, Lagrange's trigonometric identities, Euler's formula, De Moivre's formula, Morrie's law, etc.

Trigonometric identities are equations involving the trigonometric functions that are true for every value of the variables involved.

Defining relations for tangent, cotangent, secant, and cosecant in terms of sine and cosine.

$$\tan\theta = \frac{\sin\theta}{\cos\theta} \qquad \cot\theta = \frac{1}{\tan\theta} = \frac{\cos\theta}{\sin\theta}$$

$$\sec\theta = \frac{1}{\cos\theta} \qquad \csc\theta = \frac{1}{\sin\theta}$$

The Pythagorean formula for sines and cosines. This is probably the most important trig identity.

$$\sin^2\theta + \cos^2\theta = 1$$

Identities expressing trig functions in terms of their complements. Each of the six trig functions is equal to its co-function evaluated at the complementary angle.

$$\cos\theta = \sin\left(\frac{\pi}{2} - \theta\right) \quad \sin\theta = \cos\left(\frac{\pi}{2} - \theta\right)$$

$$\cot\theta = \tan\left(\frac{\pi}{2} - \theta\right) \quad \tan\theta = \cot\left(\frac{\pi}{2} - \theta\right)$$

$$\csc\theta = \sec\left(\frac{\pi}{2} - \theta\right) \quad \sec\theta = \csc\left(\frac{\pi}{2} - \theta\right)$$

Periodicity of trig functions. Sine, cosine, secant, and cosecant have period 2π while tangent and cotangent have period π.

$$\sin(\theta + 2\pi) = \sin\theta$$

$$\cos(\theta + 2\pi) = \cos\theta$$

$$\tan(\theta + \pi) = \tan\theta$$

Identities for negative angles. Sine, tangent, cotangent, and cosecant are odd functions while cosine and secant are even functions.

$$\sin(-\theta) = -\sin\theta$$
$$\cos(-\theta) = \cos\theta$$
$$\tan(-\theta) = -\tan\theta$$

Ptolemy's identities, the sum and difference formulas for sine and cosine.

$$\sin(\alpha+\beta) = \sin\alpha\cos\beta + \cos\alpha\sin\beta$$
$$\cos(\alpha+\beta) = \cos\alpha\cos\beta - \sin\alpha\sin\beta$$
$$\sin(\alpha-\beta) = \sin\alpha\cos\beta - \cos\alpha\sin\beta$$
$$\cos(\alpha-\beta) = \cos\alpha\cos\beta + \sin\alpha\sin\beta$$

Double angle formulas for sine and cosine.

$$\sin 2\theta = 2\sin\theta\cos\theta$$

$$\begin{aligned}\cos 2\theta &= \cos^2\theta - \sin^2\theta \\ &= 2\cos^2\theta - 1 \\ &= 1 - 2\sin^2\theta\end{aligned}$$

The Pythagorean formula for tangents and secants. There's also one for cotangents and cosecants, but as cotangents and cosecants are rarely needed, it's unnecessary.

$$\sec^2\theta = 1 + \tan^2\theta$$

Identities Expressing Trig Functions in terms of their Supplements.

$$\sin(\pi-\theta) = \sin\theta$$
$$\cos(\pi-\theta) = -\cos\theta$$
$$\tan(\pi-\theta) = -\tan\theta$$

Sum, Difference, and Double Angle Formulas for Tangent.

$$\tan 2\theta = \frac{2\tan\theta}{1-\tan^2\theta}$$
$$\tan(\alpha+\beta) = \frac{\tan\alpha + \tan\beta}{1-\tan\alpha\tan\beta}$$
$$\tan(\alpha-\beta) = \frac{\tan\alpha - \tan\beta}{1+\tan\alpha\tan\beta}$$

The half angle formulas. The ones for sine and cosine take the positive or negative square root depending on the quadrant of the angle $\theta/2$. For example, if $\theta/2$ is an acute angle, then the positive root would be used.

$$\sin\frac{\theta}{2} = \pm\sqrt{\frac{1-\cos\theta}{2}}$$

$$\cos\frac{\theta}{2} = \pm\sqrt{\frac{1+\cos\theta}{2}}$$

$$\tan\frac{\theta}{2} = \frac{\sin\theta}{1+\cos\theta} = \frac{1-\cos\theta}{\sin\theta}$$

Product-sum identities. This group of identities allow you to change a sum or difference of sines or cosines into a product of sines and cosines.

$$\sin\alpha + \sin\beta = 2\sin\frac{\alpha+\beta}{2}\cos\frac{\alpha-\beta}{2}$$

$$\cos\alpha + \cos\beta = 2\cos\frac{\alpha+\beta}{2}\cos\frac{\alpha-\beta}{2}$$

$$\sin\alpha - \sin\beta = 2\cos\frac{\alpha+\beta}{2}\sin\frac{\alpha-\beta}{2}$$

$$\cos\alpha - \cos\beta = -2\sin\frac{\alpha+\beta}{2}\sin\frac{\alpha-\beta}{2}$$

Product identities

$$\sin\alpha\cos\beta = \frac{\sin(\alpha+\beta)+\sin(\alpha-\beta)}{2}$$

$$\cos\alpha\cos\beta = \frac{\cos(\alpha+\beta)+\cos(\alpha-\beta)}{2}$$

$$\sin\alpha\sin\beta = \frac{\cos(\alpha-\beta)-\cos(\alpha+\beta)}{2}$$

Triple angle formulas. You can easily reconstruct these from the addition and double angle formulas.

$$\sin 3\theta = 3\sin\theta - 4\sin^3\theta$$

$$\cos 3\theta = 4\cos^3\theta - 3\cos\theta$$

$$\tan 3\theta = \frac{3\tan\theta - \tan^3\theta}{1-3\tan^2\theta}$$

More half-angle formulas. These describe the basic trig functions in terms of the tangent of half the angle. These are used in calculus for a particular kind of substitution in integrals sometimes called the Weierstrass t-substitution.

$$\sin\theta = \frac{2\tan\theta/2}{1+\tan^2\theta/2}$$

$$\cos\theta = \frac{1-\tan^2\theta/2}{1+\tan^2\theta/2}$$

$$\tan\theta = \frac{2\tan\theta/2}{1-\tan^2\theta/2}$$

Proofs of Trigonometric Identities

The eight basic trigonometric identities are listed in table below. As we will see, they are all derived from the definition of the trigonometric functions. Since many of the trigonometric identities have more than one form, we list the basic identity first and then give the most common equivalent forms.

Table:

	Basic Identities	Common Equivalent Forms
Reciprocal	$\csc\theta = \dfrac{1}{\sin\theta}$ $\sec\theta = \dfrac{1}{\cos\theta}$ $\cot\theta = \dfrac{1}{\tan\theta}$	$\sin\theta = \dfrac{1}{\csc\theta}$ $\cos\theta = \dfrac{1}{\sec\theta}$ $\tan\theta = \dfrac{1}{\cot\theta}$
Ratio	$\tan\theta = \dfrac{\sin\theta}{\cos\theta}$ $\cot\theta = \dfrac{\cos\theta}{\sin\theta}$	
Pythagorean	$\cos^2\theta + \sin^2\theta = 1$ $1 + \tan^2\theta = \sec^2\theta$ $1 + \cot^2\theta = \csc^2\theta$	$\sin^2\theta = 1 - \cos^2\theta$ $\sin\theta = \pm\sqrt{1 - \cos^2\theta}$ $\cos^2\theta = 1 - \sin^2\theta$ $\cos\theta = \pm\sqrt{1 - \sin^2\theta}$

Reciprocal Identities

Note that, in above Table, the eight basic identities are grouped in categories. For example, since $\csc\ \theta = 1/(\sin\ \theta)$, cosecant and sine must be reciprocals. It is for this reason that we call the identities in this category. reciprocal identities.

As we mentioned above, the eight basic identities are all derived from the definition of the six trigonometric functions. To derive the first reciprocal identity, we use the definition of $\sin\ \theta$ to write

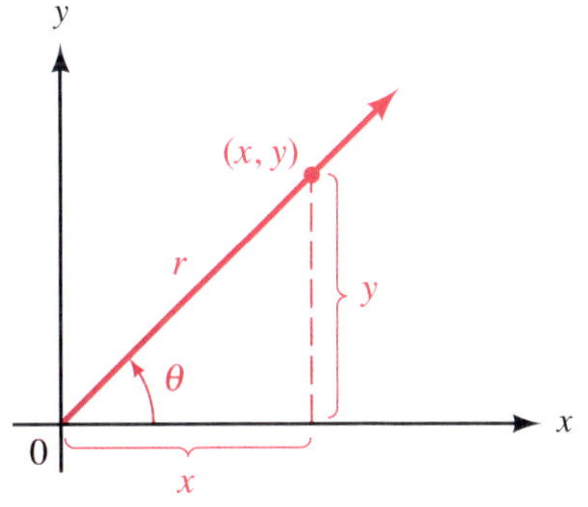

$$\frac{1}{\sin\theta} = \frac{1}{y/r} = \frac{r}{y} = \csc\theta$$

Note that we can write this same relationship between $\sin\ \theta$ and $\csc\theta$ as

$$\sin\theta = \frac{1}{\csc\theta}$$

Because

$$\frac{1}{\csc\theta} = \frac{1}{r/y} = \frac{y}{r} = \sin\theta$$

The first identity we wrote, $\csc\theta = 1/(\sin\ \theta)$, is the basic identity. The second one, $\sin\theta = 1/(\csc\theta)$, is an equivalent form of the first one. The other reciprocal identities and their common equivalent forms are derived in a similar manner.

Ratio Identities

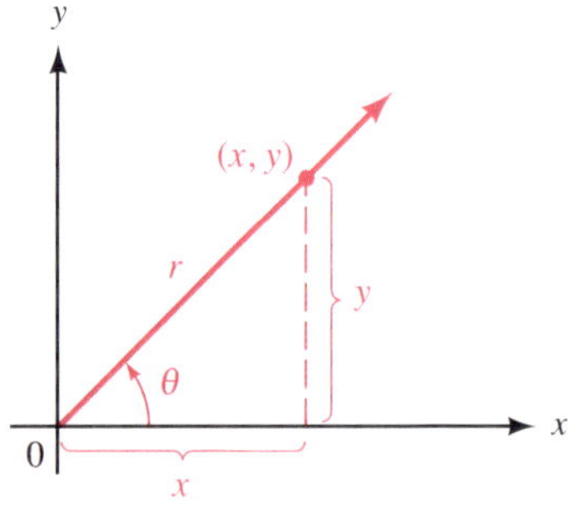

Unlike the reciprocal identities, the ratio identities do not have any common equivalent forms. Here is how we derive the ratio identity for $\tan\theta$:

$$\frac{\sin\theta}{\cos\theta} = \frac{y/r}{x/r} = \frac{y}{x} = \tan\theta$$

Example: If $\sin\theta = -\frac{3}{5}$ and $\cos\theta = \frac{4}{5}$, find $\tan\theta$ and $\cot\theta$.

Solution: Using the ratio identities we have

$$\tan\theta = \frac{\sin\theta}{\cos\theta} = \frac{-\frac{3}{5}}{\frac{4}{5}} = -\frac{3}{4}$$

$$\cot\theta = \frac{\cos\theta}{\sin\theta} = \frac{\frac{4}{5}}{-\frac{3}{5}} = -\frac{4}{3}$$

Note that, once we found $\tan\theta$, we could have used a reciprocal identity to find $\cot\theta$

$$\cot\theta = \frac{1}{\tan\theta} = \frac{1}{-\frac{3}{4}} = -\frac{4}{3}$$

Pythagorean Identities

The identity $\cos^2\theta + \sin^2\theta = 1$ is called a Pythagorean identity because it is derived from the Pythagorean Theorem. Recall from the definition of $\sin\theta$ and $\cos\theta$ that if (x, y) is a point on the terminal side of θ and r is the distance to (x, y) from the origin, the relationship between x, y, and r is $x^2 + y^2 = r^2$. This relationship comes from the Pythagorean Theorem. Here is how we use it to derive the first Pythagorean identity.

$$x^2 + y^2 = r^2$$

$$\frac{x^2}{r^2} + \frac{y^2}{r^2} = 1 \qquad \text{Divide each side by } r^2.$$

$$\left(\frac{x}{r}\right)^2 + \left(\frac{y}{r}\right)^2 = 1 \qquad \text{Property of exponents.}$$

$$(\cos\theta)^2 + (\sin\theta)^2 = 1 \qquad \text{Definition of } \sin\theta \text{ and } \cos\theta$$

$$\cos^2\theta + \sin^2\theta = 1 \qquad \text{Notation}$$

There are four very useful equivalent forms of the first Pythagorean identity. Two of the forms occur when we solve $\cos^2\theta + \sin^2\theta = 1$ for $\cos\theta$ while the other two forms are the result of solving for $\sin\theta$.

Solving $\cos^2\theta + \sin^2\theta = 1$ for $\cos\theta$ we have

$$\cos^2\theta + \sin^2\theta = 1$$

$$\cos^2\theta = 1 - \sin^2\theta \qquad \text{Add} - \sin^2\theta \text{ to each side.}$$

$$\cos\ \theta = \pm\sqrt{1 - \sin^2\theta} \qquad \text{Take the square root of each side.}$$

Similarly, solving for $\sin\theta$ gives us

$$\sin^2\theta = 1 - \cos^2\theta$$

And

$$\sin\ \theta = \pm\sqrt{1 - \cos^2\theta}$$

Example: If $\sin\theta = \frac{3}{5}$ and θ terminates in quadrant II, find $\cos\theta$

Solution: We can obtain $\cos\theta$ from $\sin\ \theta$ by using the identity

$$\cos\theta = \pm\sqrt{1-\sin^2\theta}$$

If $\sin\theta = \frac{3}{5}$, the identity becomes

$\cos\theta = \pm\sqrt{1-\left(\frac{3}{5}\right)^2}$ Substitute $\frac{3}{5}$ for $\sin\theta$.

$= \pm\sqrt{1-\left(\frac{9}{25}\right)}$ Square $\frac{3}{5}$ to get $\frac{9}{25}$

$= \pm\sqrt{\left(\frac{16}{25}\right)}$ Subtract.

$= \pm\frac{4}{5}$ Take the square root of the numerator and denominator separately.

Now we know that $\cos\theta$ is either $+\frac{4}{5}$ or $-\frac{4}{5}$ Looking back to the original statement of the problem, however, we see that θ terminates in quadrant II; therefore, $\cos\theta$ must be negative.

$$\cos\theta = -\frac{4}{5}$$

Example: If $\cos\theta$ — and θ terminates in quadrant IV, find the remaining trigonometric ratios for θ.

Solution: The first, and easiest, ratio to find is $\sec\theta$, because it is the reciprocal of $\cos\theta$.

$$\sec\theta = \frac{1}{\cos\theta} = \frac{1}{\frac{1}{2}} = 2$$

Next, we find $\sin\theta$. Since θ terminates in QIV $\sin\theta$ will be negative. Using one of the equivalent forms of the Pythagorean identity, we have

$\sin\theta = -\sqrt{1-\cos^2\theta}$ Negative sign because θ is in QIV.

$= -\sqrt{1-\left(\frac{1}{2}\right)^2}$ Substitute $\frac{1}{2}$ for $\cos\theta$.

$= -\sqrt{1-\frac{1}{4}}$ Square $\frac{1}{2}$ to get $\frac{1}{4}$

$= -\sqrt{\frac{3}{4}}$ Subtract.

$= -\frac{\sqrt{3}}{2}$ Take the square root of the numerator and denominator separately.

Now that we have $\sin\theta$ and $\cos\theta$ we can find $\tan\theta$ by using a ratio identity.

$$\tan\theta = \frac{\sin\theta}{\cos\theta} = \frac{-\sqrt{3}/2}{1/2} = -\sqrt{3}$$

$\cot\theta$ and $\csc\theta$ are the reciprocals of $\tan\theta$ and $\sin\theta$, respectively. Therefore,

$$\cot\theta = \frac{1}{\tan\theta} = -\frac{1}{\sqrt{3}} \qquad \csc\theta = \frac{1}{\sin\theta} = -\frac{2}{\sqrt{3}}$$

Here are all six ratios together:

$$\sin\theta = -\frac{\sqrt{3}}{2} \qquad \csc\theta = -\frac{2}{\sqrt{3}}$$

$$\cos\theta = \frac{1}{2} \qquad \sec\theta = 2$$

$$\tan\theta = -\sqrt{3} \qquad \cot\theta = -\frac{1}{\sqrt{3}}$$

The basic identities allow us to write any of the trigonometric functions in terms of sine and cosine. The next examples illustrate this.

Example: Write $\tan\theta$ in terms of $\sin\theta$

Solution When we say we want $\tan\theta$ written in terms of $\sin\theta$ we mean that we want to write an expression that is equivalent to $\tan\theta$ but involves no trigonometric function other than $\sin\theta$.Let's begin by using a ratio identity to write tan in terms of $\sin\theta$ and $\cos\theta$:

$$\tan\theta = \frac{\sin\theta}{\cos\theta}$$

Now we need to replace $\cos\theta$ with an expression involving only $\sin\theta$.Since $\cos\theta = \pm\sqrt{1-\sin^2\theta}$, we have

$$\begin{aligned}\tan\theta &= \frac{\sin\theta}{\cos\theta}\\ &= \frac{\sin\theta}{\pm\sqrt{1-\sin^2\theta}}\\ &= \pm\frac{\sin\theta}{\sqrt{1-\sin^2\theta}}\end{aligned}$$

This last expression is equivalent to $\tan\theta$ and is written in terms of $\sin\theta$ only. (In a problem like this it is okay to include numbers and algebraic symbols with $\sin\theta$ — just no other trigonometric functions.)

Here is another example. This one involves simplification of the product of two trigonometric functions.

Example: Write $\sec\theta\tan\theta$ in terms of $\sin\theta$ and $\cos\theta$, and then simplify.

Solution: Since $\sec\theta = 1/(\cos\theta)$ and $\tan\theta = (\sin\theta)/(\cos\theta)$, we have

$$\begin{aligned}\sec\theta\tan\theta &= \frac{1}{\cos\theta}\cdot\frac{\sin\theta}{\cos\theta}\\ &= \frac{\sin\theta}{\cos^2\theta}\end{aligned}$$

The next examples show how we manipulate trigonometric expressions using algebraic techniques.

Example: Add $= \frac{1}{\sin\theta} + \frac{1}{\cos\theta}$.

Solution: We can add these two expressions in the same way we would add $\frac{1}{3}$ and $\frac{1}{4}$ by first finding a least common denominator, and then writing each expression again with the LCD for its denominator.

$$\begin{aligned}\frac{1}{\sin\theta} + \frac{1}{\cos\theta} &= \frac{1}{\sin\theta}\cdot\frac{\cos\theta}{\cos\theta} + \frac{1}{\cos\theta}\cdot\frac{\sin\theta}{\sin\theta} \quad \text{The LCD is } \sin\theta\cos\theta.\\ &= \frac{\cos\theta}{\sin\theta\cos\theta} + \frac{\sin\theta}{\cos\theta\sin\theta}\\ &= \frac{\cos\theta + \sin\theta}{\sin\theta\cos\theta}\end{aligned}$$

Example: Multiply $(\sin\theta + 2)(\sin\theta - 5)$.

Solution: We multiply these two expressions in the same way we would multiply $(x+2)(x-5)$.

$$\begin{aligned}&\qquad\qquad\qquad\quad \text{F} \qquad\qquad \text{O} \qquad\quad \text{I} \qquad \text{L}\\ (\sin\theta+2)(\sin\theta-5) &= \sin\theta\,\sin\theta - 5\ \sin\theta + 2\ \sin\theta - 10\\ &= \sin^2\theta - 3\sin\theta - 10\end{aligned}$$

Proving Identities

Next we want to use the eight basic identities and their equivalent forms to verify other trigonometric identities. To prove (or verify) that a trigonometric identity is true, we use trigonometric substitutions and algebraic manipulations to either:

1. Transform the right side into the left side.

Or:

2. Transform the left side into the right side.

The main thing to remember in proving identities is to work on each side of the identity separately. We do not want to use properties from algebra that involve both sides of the identity—such as the addition property of equality. We prove identities in order to develop the ability to transform one trigonometric expression into another. When we encounter problems in other courses that require the use of the techniques used to verify identities, we usually find that the solution to these problems hinges upon transforming an expression containing trigonometric functions into a less complicated expression. In these cases, we do not usually have an equal sign to work with.

Example: Verify the identity: $\sin\theta \cot\theta = \cos\theta$.

Proof to prove this identity we transform the left side into the right side:

$$\sin\theta\cot\theta = \sin\theta.\frac{\cos\theta}{\sin\theta} \quad \text{Ratio identity}$$

$$= \frac{\sin\theta\cos\theta}{\sin\theta} \quad \text{Multiply.}$$

$$= \cos\theta \quad \textit{Divide out common factor } \sin\theta.$$

Example: prove : $\tan x + \cos x = \sin x(\sec x + \cot x)$.

Proof : We begin by applying the distributive property to the right side of the identity. Then we change each expression on the right side to an equivalent expression involving only sin x and cos x.

$$\sin x(\sec x + \cot x) = \sin x \sec x + \sin x \cot x \quad \text{Multiply.}$$

$$= \sin x.\frac{1}{\cos x} + \sin x.\frac{\cos x}{\sin x} \quad \text{Reciprocal and ratio identities}$$

$$= \frac{\sin x}{\cos x} + \cos x \quad \text{Multiply and divide out common factor } \sin x$$

$$= \tan x + \cos x \quad \text{Ratio identity}$$

In this case, we transformed the right side into the left side.

Example

Prove: $\frac{\cos^4 t - \sin^4 t}{\cos^2 t} = 1 - \tan^2 t.$

Proof In this example, factoring the numerator on the left side will reduce the exponents there from 4 to 2.

$$\frac{\cos^4 t - \sin^4 t}{\cos^2 t} = \frac{(\cos^2 t + \sin^2 t)(\cos^2 t - \sin^2 t)}{\cos^2 t} \quad \text{Factor.}$$

$$= \frac{1(\cos^2 t - \sin^2 t)}{\cos^2 t} \quad \text{Pythagorean identity}$$

$$= \frac{\cos^2 t}{\cos^2 t} - \frac{\sin^2 t}{\cos^2 t} \quad \text{Separate into two fractions.}$$

$$= 1 - \tan^2 t \quad \text{Ratio identity}$$

Example

Prove: $1 + \cos\theta = \frac{\sin^2\theta}{1 - \cos\theta}.$

Proof: We begin by applying an alternative form of the Pythagorean identity to the right side to write $\sin^2\theta$ as $1-\cos^2\theta$. Then we factor $1-\cos^2\theta$ and reduce to lowest terms.

$$\frac{\sin^2\theta}{1-\cos\theta}=\frac{1-\cos^2\theta}{1-\cos\theta} \quad \text{Pythagorean identity}$$

$$=\frac{(1-\cos\theta)(1+\cos\theta)}{1-\cos\theta} \quad \text{Factor.}$$

$$=1+\cos\theta \quad \text{Reduce.}$$

Example

Prove: $\tan x+\cot x=\sec x\csc x$.

Proof: We begin by rewriting the left side in terms of $\sin x$ and $\cos x$. Then we simplify by finding a common denominator, changing to equivalent fractions, and adding them.

$$\tan x+\cot x=\frac{\sin x}{\cos x}+\frac{\cos x}{\sin x} \quad \text{Change to expressions in } \sin x \text{ and } \cos x$$

$$=\frac{\sin x}{\cos x}\cdot\frac{\sin x}{\sin x}+\frac{\cos x}{\sin x}\cdot\frac{\cos x}{\cos x} \quad \text{LCD}$$

$$=\frac{\sin^2 x+\cos^2 x}{\cos x\sin x} \quad \text{Add fractions.}$$

$$=\frac{1}{\cos x\sin x} \quad \text{Pythagorean identity}$$

$$=\frac{1}{\cos x}\cdot\frac{1}{\sin x} \quad \text{Write as separate fractions.}$$

$$=\sec x\csc x \quad \text{Reciprocal identities}$$

Example

Prove: $\frac{\sin A}{1+\cos A}+\frac{1+\cos A}{\sin A}=2\csc A$.

Proof : The LCD for the left side is $\sin A(1+\cos A)$.

$$\frac{\sin A}{1+\cos A}+\frac{1+\cos A}{\sin A}=\frac{\sin A}{\sin A}\cdot\frac{\sin A}{1+\cos A}+\frac{1+\cos A}{\sin A}\cdot\frac{1+\cos A}{1+\cos A} \quad \text{LCD}$$

$$=\frac{\sin^2 A+(1+\cos A)^2}{\sin A+(1+\cos A)} \quad \text{Add fractions.}$$

$$=\frac{\sin^2 A+1+2\cos A+\cos^2 A}{\sin A+(1+\cos A)} \quad \text{Expand } (1+\cos A)^2.$$

$$=\frac{2+2\cos A}{\sin A(1+\cos A)} \quad \textit{Pythagorean identity}$$

$$=\frac{2(1+\cos A)}{\sin A(1+\cos A)} \quad \text{Factor out 2.}$$

$$= \frac{2}{\sin A} \qquad \text{Reduce.}$$

$$= 2\csc A \qquad \text{Reciprocal identity}$$

Example

Prove: $\frac{1+\sin t}{\cos t} = \frac{\cos t}{1-\sin t}$.

Proof: The trick to proving this identity is to multiply the numerator and denominator on the right side by $1+\sin t$

$$\frac{\cos t}{1-\sin t} = \frac{\cos t}{1-\sin t} \cdot \frac{1+\sin t}{1+\sin t} \qquad \text{Multiply numerator and denominator by } 1+\sin t.$$

$$= \frac{\cos t(1+\sin t)}{1-\sin^2 t} \qquad \text{Multiply out the denominator.}$$

$$= \frac{\cos t(1+\sin t)}{\cos^2 t} \qquad \text{Pythagorean identity}$$

$$= \frac{(1+\sin t)}{\cos t} \qquad \text{Reduce}$$

Note that it would have been just as easy for us to verify this identity by multiplying the numerator and denominator on the left side by $1-\sin t$.

Sum and Difference Formulas

The expressions $\sin(A+B)$ and $\cos(A+B)$ occur frequently enough in mathematics that it is necessary to find expressions equivalent to them that involve sines and cosines of single angles. The most obvious question to begin with is

$$\sin(A+B) = \sin A + \sin B?$$

The answer is no. Substituting almost any pair of numbers for A and B in the formula will yield a false statement. As a counterexample, we can let $A = 30^\circ$ and $B = 60^\circ$ in the formula above and then simplify each side.

$$\sin(30^\circ + 60^\circ) \stackrel{?}{=} \sin 30^\circ + \sin 60^\circ$$

$$\sin 90^\circ \stackrel{?}{=} \frac{1}{2} + \frac{\sqrt{3}}{2}$$

$$1 \neq \frac{1+\sqrt{3}}{2}$$

The formula just doesn't work. The next question is, what are the formulas for $\sin(A+B)$ and $\cos(A+B)$? The answer to that question is what this section is all about. Let's start by deriving the formula for $\cos(A+B)$.

We begin by drawing A in standard position and then adding B and B to it. These angles are shown in below figure in relation to the unit circle. The unit circle is the circle with its center at the origin and with a radius of 1. Since the radius of the unit circle is 1, the point through which the terminal side of A passes will have coordinates $(\cos A, \sin A)$. [If P_2 in Figure has coordinates (x, y), then by the definition of sin A, cos A, and the unit circle, $\cos A = x/r = x/1 = x$ and $\sin A = y/r = y/1 =$. Therefore, $(x, y) = (\cos A, \sin A)$.] The points on the unit circle through which the terminal sides of the other angles in Figure pass are found in the same manner.

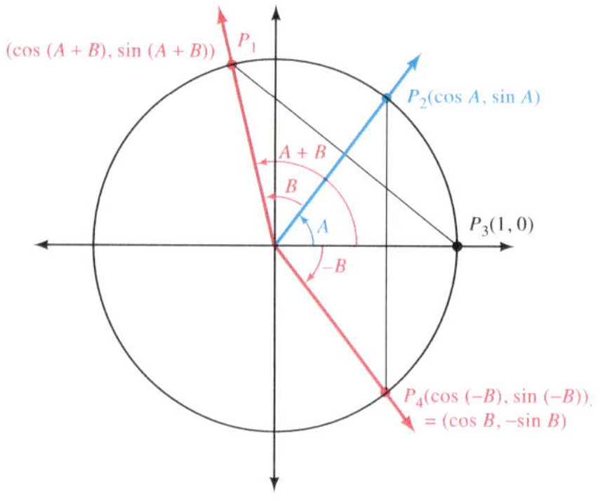

To derive the formula for $(\cos A + B)$, we simply have to see that line segment P_1P_3 is equal to line segment P_2P_4. (From geometry, they are chords cut off by equal central angles.)

$$\overline{P_1P_3} = \overline{P_2P_4}$$

Squaring both sides gives us

$$\left(\overline{P_1P_3}\right)^2 = \left(\overline{P_2P_4}\right)^2$$

Now, applying the distance formula, we have

$$\left[\cos(A+B) - 1\right]^2 = \left[\sin(A+B) - 0\right]^2 = (\cos A - \cos B)^2 + (\sin A + \sin B)^2$$

Let's call this Equation. Taking the left side of Equation 1, expanding it, and then simplifying by using the Pythagorean identity, gives us

Left side of Equation.

$$\cos^2(A+B) - 2\cos(A+B) + 1 + \sin^2(A+B) \qquad \text{Expand squares.}$$

$$= -2\cos(A+B) + 2 \qquad \text{Pythagorean identity}$$

Applying the same two steps to the right side of Equation looks like this:

Right side of Equation.

$$\cos^2 A - 2\cos A\cos B + \cos^2 B + \sin^2 A + 2\sin A\sin B + \sin^2 B$$

$$= -2\cos A\cos B + 2\sin A\sin B + 2$$

Equating the simplified versions of the left and right sides of Equation, we have

$$-2\cos(A+B)+2 = -2\cos A\cos B+2\sin A\sin B+2$$

Adding -2 to both sides and then dividing both sides by -2 gives us the formula we are after.

$$\cos(A+B)=\cos A\cos B-\sin A\sin B$$

This is the first formula in a series of formulas for trigonometric functions of the sum or difference of two angles. It must be memorized. Before we derive the others, let's look at some of the ways we can use our first formula.

Example

Find the exact value for cos 75°.

Solution We write 75° as $45^\circ+30^\circ$ and then apply the formula for $\cos(A+B)$.

$$\begin{aligned}\cos 75^\circ &= \cos\left(45^\circ+30^\circ\right)\\ &= \cos 45^\circ\cos 30^\circ-\sin 45^\circ\sin 30^\circ\\ &= \frac{\sqrt{2}}{2}.\frac{\sqrt{3}}{2}-\frac{\sqrt{2}}{2}.\frac{1}{2}\\ &= \frac{\sqrt{6}-\sqrt{2}}{4}\end{aligned}$$

Tangent Half-angle Formula

Use the ratio identity for tangent and fill in the half-angle identities for sine and cosine.

$$\tan\frac{\theta}{2}=\frac{\sin\frac{\theta}{2}}{\cos\frac{\theta}{2}}=\frac{\sqrt{\frac{1-\cos\theta}{2}}}{\sqrt{\frac{1+\cos\theta}{2}}}$$

You can leave off the ± sign because you won't have to choose which sign to use with the tangent identity — the square-root sign squares out of the equation.

Put the numerator and denominator under the same radical and then simplify the complex fraction.

$$=\frac{\sqrt{\frac{1-\cos\theta}{2}}}{\sqrt{\frac{1+\cos\theta}{2}}}=\sqrt{\frac{1-\cos\theta}{2}.\frac{2}{1+\cos\theta}}=\sqrt{\frac{1-\cos\theta}{1+\cos\theta}}$$

Multiply the numerator and denominator by the conjugate (same terms, different sign) of the denominator.

$$\sqrt{\frac{1-\cos\theta}{1+\cos\theta}\cdot\frac{1-\cos\theta}{1-\cos\theta}}=\sqrt{\frac{(1-\cos\theta)^2}{1+\cos^2\theta}}$$

Replace the denominator by using the Pythagorean identity and then simplify by putting the radical over the numerator and denominator, individually.

$$\sqrt{\frac{(1-\cos\theta)^2}{1-\cos^2\theta}}=\sqrt{\frac{(1-\cos\theta)^2}{\sin^2\theta}}$$

$$=\frac{\sqrt{(1-\cos\theta)^2}}{\sqrt{\sin^2\theta}}=\frac{1-\cos\theta}{\sin\theta}$$

To find the other form of this half-angle tangent identity, change Step 3 by multiplying the numerator and denominator of the fraction by the conjugate of the *numerator* instead of the denominator.

$$\tan\left(\frac{\eta\pm\theta}{2}\right)=\frac{\sin\eta\pm\sin\theta}{\cos\eta+\cos\theta}=-\frac{\cos\eta-\cos\theta}{\sin\eta\mp\sin\theta},$$

$$\tan\left(\pm\frac{\theta}{2}\right)=\frac{\pm\sin\theta}{1+\cos\theta}=\frac{\pm\tan\theta}{\sec\theta+1}=\frac{\pm1}{\csc\theta+\cot\theta},\qquad(\eta=0)$$

$$\tan\left(\pm\frac{\theta}{2}\right)=\frac{1-\cos\theta}{\pm\sin\theta}=\frac{\sec\theta-1}{\pm\tan\theta}=\pm(\csc\theta-\cot\theta),\qquad(\eta=0)$$

$$\tan\left(\frac{1}{2}(\theta\pm\frac{\pi}{2})\right)=\frac{1\pm\sin\theta}{\cos\theta}=\sec\theta\pm\tan\theta=\frac{\csc\theta\pm1}{\cot\theta},\qquad(\eta=\frac{\pi}{2})$$

$$\tan\left(\frac{1}{2}(\theta\pm\frac{\pi}{2})\right)=\frac{\cos\theta}{1\mp\sin\theta}=\frac{1}{\sec\theta\mp\tan\theta}=\frac{\cot\theta}{\csc\theta\mp1},\qquad(\eta=\frac{\pi}{2})$$

$$\frac{1-\tan(\theta/2)}{1+\tan(\theta/2)}=\pm\sqrt{\frac{1-\sin\theta}{1+\sin\theta}}$$

$$\tan\frac{\theta}{2}=\pm\sqrt{\frac{1-\cos\theta}{1+\cos\theta}}$$

From these one can derive identities expressing the sine, cosine, and tangent as functions of tangents of half-angles:

$$\sin\alpha=\frac{2\tan\frac{\alpha}{2}}{1+\tan^2\frac{\alpha}{2}}$$

$$\cos\alpha=\frac{1-\tan^2\frac{\alpha}{2}}{1+\tan^2\frac{\alpha}{2}}$$

$$\tan\alpha = \frac{2\tan\frac{\alpha}{2}}{1-\tan^2\frac{\alpha}{2}}$$

Proofs

Algebraic Proofs

Use double-angle formulae and $\sin^2\alpha + \cos^2\alpha = 1$,

$$\sin\alpha = 2\sin\frac{\alpha}{2}\cos\frac{\alpha}{2} = \frac{2\sin\frac{\alpha}{2}\cos\frac{\alpha}{2}}{\cos^2\frac{\alpha}{2}+\sin^2\frac{\alpha}{2}} = \frac{2\tan\frac{\alpha}{2}}{1+\tan^2\frac{\alpha}{2}}$$

$$\cos\alpha = \cos^2\frac{\alpha}{2} - \sin^2\frac{\alpha}{2} = \frac{\cos^2\frac{\alpha}{2}-\sin^2\frac{\alpha}{2}}{\cos^2\frac{\alpha}{2}+\sin^2\frac{\alpha}{2}} = \frac{\frac{\cos^2\frac{\alpha}{2}}{\cos^2\frac{\alpha}{2}} - \frac{\sin^2\frac{\alpha}{2}}{\cos^2\frac{\alpha}{2}}}{\frac{\cos^2\frac{\alpha}{2}}{\cos^2\frac{\alpha}{2}} + \frac{\sin^2\frac{\alpha}{2}}{\cos^2\frac{\alpha}{2}}} = \frac{1-\tan^2\frac{\alpha}{2}}{1+\tan^2\frac{\alpha}{2}}$$

taking the quotient of the formulae for sine and cosine yields

$$\tan\alpha = \frac{2\tan\frac{\alpha}{2}}{1-\tan^2\frac{\alpha}{2}}$$

Combining the Pythagorean identity $\cos^2\alpha + \sin^2\alpha = 1$ with the double-angle formula for the cosine, $\cos 2\alpha = \cos^2\alpha - \sin^2\alpha = 1 - 2\sin^2\alpha = 2\cos^2\alpha - 1$,

rearranging, and taking the square roots yields

$$|\sin\alpha| = \sqrt{\frac{1-\cos 2\alpha}{2}} \text{ and } |\cos\alpha| = \sqrt{\frac{1+\cos 2\alpha}{2}}$$

which, upon division gives

$$|\tan\alpha| = \frac{\sqrt{1-\cos 2\alpha}}{\sqrt{1+\cos 2\alpha}} = \frac{\sqrt{1-\cos 2\alpha}\sqrt{1+\cos 2\alpha}}{1+\cos 2\alpha} = \frac{\sqrt{1-\cos^2 2\alpha}}{1+\cos 2\alpha} = \frac{|\sin 2\alpha|}{1+\cos 2\alpha}$$

or alternatively

$$|\tan\alpha| = \frac{\sqrt{1-\cos 2\alpha}}{\sqrt{1+\cos 2\alpha}} = \frac{1-\cos 2\alpha}{\sqrt{1+\cos 2\alpha}\sqrt{1-\cos 2\alpha}} = \frac{1-\cos 2\alpha}{\sqrt{1-\cos^2 2\alpha}} = \frac{1-\cos 2\alpha}{|\sin 2\alpha|}.$$

Also, using the angle addition and subtraction formulae for both the sine and cosine one obtains:

$$\cos(a+b) = \cos a \cos b - \sin a \sin b$$

$$\cos(a-b) = \cos a \cos b + \sin a \sin b$$

$$\sin(a+b) = \sin a \cos b + \cos a \sin b$$

$$\sin(a-b) = \sin a \cos b - \cos a \sin b$$

Pairwise addition of the above four formulae yields:

$$\begin{aligned}\sin(a+b)+\sin(a-b) &= \sin a \cos b + \cos a \sin b + \sin a \cos b - \cos a \sin b \\ &= 2\sin a \cos b \\ \cos(a+b)+\cos(a-b) &= \cos a \cos b - \sin a \sin b + \cos a \cos b + \sin a \sin b \\ &= 2\cos a \cos b\end{aligned}$$

Setting $a = \dfrac{p+q}{2}$ and $b = \dfrac{p-q}{2}$ substituting yields:

$$\begin{aligned}\sin\left(\frac{p+q}{2}+\frac{p-q}{2}\right)+\sin\left(\frac{p+q}{2}-\frac{p-q}{2}\right) &= \sin(p)+\sin(q) \\ &= 2\sin\left(\frac{p+q}{2}\right)\cos\left(\frac{p-q}{2}\right) \\ \cos\left(\frac{p+q}{2}+\frac{p-q}{2}\right)+\cos\left(\frac{p+q}{2}-\frac{p-q}{2}\right) &= \cos(p)+\cos(q) \\ &= 2\cos\left(\frac{p+q}{2}\right)\cos\left(\frac{p-q}{2}\right)\end{aligned}$$

Dividing the sum of sines by the sum of cosines one arrives at:

$$\frac{\sin(p)+\sin(q)}{\cos(p)+\cos(q)} = \frac{2\sin\left(\frac{p+q}{2}\right)\cos\left(\frac{p-q}{2}\right)}{2\cos\left(\frac{p+q}{2}\right)\cos\left(\frac{p-q}{2}\right)} = \tan\left(\frac{p+q}{2}\right)$$

Geometric Proofs

Applying the formulae derived above to the rhombus figure on the right, it is readily shown that

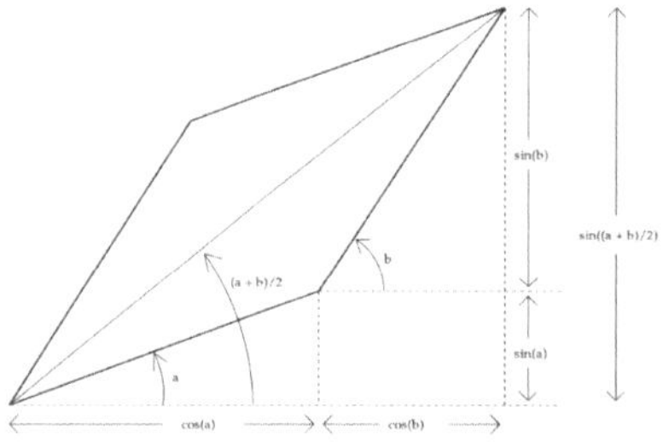

The sides of this rhombus have length 1. The angle between the horizontal line and the shown diagonal is $(a+b)/2$. This is a geometric way to prove a tangent half-angle formula. Note that $\sin((a+b)/2)$ and $\cos((a+b)/2)$ just show their relation to the diagonal, not the real value.

$$\tan\frac{a+b}{2} = \frac{\sin\frac{a+b}{2}}{\cos\frac{a+b}{2}} = \frac{\sin a + \sin b}{\cos a + \cos b}.$$

In the unit circle, application of the above shows that $t = \tan\left(\frac{\varphi}{2}\right)$. According to similar triangles,

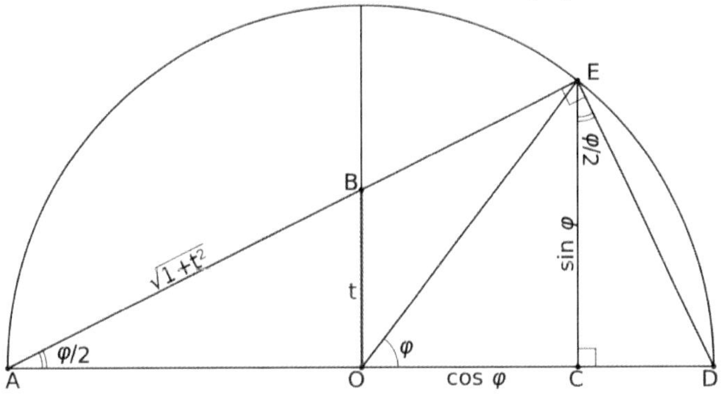

A geometric proof of the tangent half-angle formula

$\frac{t}{\sin\varphi} = \frac{1}{1+\cos\varphi}$. It follows that

$$t = \frac{\sin\varphi}{1+\cos\varphi} = \frac{\sin\varphi(1-\cos\varphi)}{(1+\cos\varphi)(1-\cos\varphi)} = \frac{1-\cos\varphi}{\sin\varphi}.$$

The tangent Half-angle Substitution in Integral Calculus

In various applications of trigonometry, it is useful to rewrite the trigonometric functions (such as sine and cosine) in terms of rational functions of a new variable t. These identities are known collectively as the tangent half-angle formulae because of the definition of t. These identities can be useful in calculus for converting rational functions in sine and cosine to functions of t in order to find their antiderivatives.

Technically, the existence of the tangent half-angle formulae stems from the fact that the circle is an algebraic curve of genus 0. One then expects that the *circular functions* should be reducible to rational functions.

Geometrically, the construction goes like this: for any point $(\cos \varphi, \sin \varphi)$ on the unit circle, draw the line passing through it and the point $(-1, 0)$. This point crosses the y-axis at some point $y = t$. One can show using simple geometry that $t = \tan(\varphi/2)$. The equation for the drawn line is $y = (1 + x)t$. The equation for the intersection of the line and circle is then a quadratic equation involving t. The two solutions to this equation are $(-1, 0)$ and $(\cos \varphi, \sin \varphi)$. This allows us to write the latter as rational functions of t (solutions are given below).

Note also that the parameter t represents the stereographic projection of the point $(\cos \varphi, \sin \varphi)$ onto the y-axis with the center of projection at $(-1, 0)$. Thus, the tangent half-angle formulae give conversions between the stereographic coordinate t on the unit circle and the standard angular coordinate φ.

Then we have

$$\cos\varphi = \frac{1-t^2}{1+t^2}, \quad \sin\varphi = \frac{2t}{1+t^2},$$

$$\tan\varphi = \frac{2t}{1-t^2} \quad \cot\varphi = \frac{1-t^2}{2t},$$

$$\sec\varphi = \frac{1+t^2}{1-t^2}, \quad \csc\varphi = \frac{1+t^2}{2t},$$

and

$$e^{i\varphi} = \frac{1+it}{1-it}, \qquad e^{-i\varphi} = \frac{1-it}{1+it}.$$

By eliminating phi between the directly above and the initial definition of t, one arrives at the following useful relationship for the arctangent in terms of the natural logarithm

$$\arctan t = \frac{1}{2i}\ln\frac{1+it}{1-it}.$$

In calculus, the Weierstrass substitution is used to find antiderivatives of rational functions of sin φ and cos φ. After setting

$$t = \tan\frac{1}{2}\varphi.$$

This implies that

$$\varphi = 2\arctan(t) + 2\pi n,$$

for some integer n, and therefore

$$d\varphi = \frac{2dt}{1+t^2}.$$

Hyperbolic Identities

One can play an entirely analogous game with the hyperbolic functions. A point on (the right branch of) a hyperbola is given by (cosh θ, sinh θ). Projecting this onto y-axis from the center (−1, 0) gives the following:

$$t = \tanh\frac{1}{2}\theta = \frac{\sinh\theta}{\cosh\theta + 1} = \frac{\cosh\theta - 1}{\sinh\theta}$$

with the identities

$$\cosh\,\theta = \frac{1+t^2}{1-t^2}, \quad \sinh\,\theta = \frac{2t}{1-t^2},$$

$$\tanh \theta = \frac{2t}{1+t^2}, \quad \coth \theta = \frac{1+t^2}{2t},$$

$$\operatorname{sech} \theta = \frac{1-t^2}{1+t^2}, \quad \operatorname{csch} \theta = \frac{1-t^2}{2t},$$

and

$$e^{\theta} = \frac{1+t}{1-t}, \qquad e^{-\theta} = \frac{1-t}{1+t}.$$

The use of this substitution for finding antiderivatives was introduced by Karl Weierstrass.

Finding θ in terms of t leads to following relationship between the hyperbolic arctangent and the natural logarithm:

$$\operatorname{artanh} t = \frac{1}{2} \ln \frac{1+t}{1-t}.$$

("ar-" is used rather than "arc-" because "arc" is about arc length and "ar" abbreviates "area". It is the area between two rays and a hyperbola, rather than the arc length between two rays measure along an arc a circle.)

The Gudermannian Function

Comparing the hyperbolic identities to the circular ones, one notices that they involve the same functions of t, just permuted. If we identify the parameter t in both cases we arrive at a relationship between the circular functions and the hyperbolic ones. That is, if

$$t - \tan \frac{1}{2} \varphi - \tanh \frac{1}{2} \theta$$

then

$$\varphi = 2 \tan^{-1} \tanh \frac{1}{2} \theta \equiv \operatorname{gd} \theta.$$

where $\operatorname{gd}(\theta)$ is the Gudermannian function. The Gudermannian function gives a direct relationship between the circular functions and the hyperbolic ones that does not involve complex numbers. The above descriptions of the tangent half-angle formulae (projection the unit circle and standard hyperbola onto the y-axis) give a geometric interpretation of this function.

Pythagorean Triples

The tangent of half of an acute angle of a right triangle whose sides are a Pythagorean triple will necessarily be a rational number in the interval (0, 1). Vice versa, when a half-angle tangent is a rational number in the interval (0, 1), there is a right triangle that has the full angle and that has side lengths that are a Pythagorean triple.

Lagrange's Trigonometric Identities

In trigonometry, Lagrange's trigonometric identities, named after Joseph Louis Lagrange, are:

$$\sum_{n=1}^{N} \sin n\theta = \frac{1}{2}\cot\theta - \frac{\cos\left(N+\frac{1}{2}\right)\theta}{2\sin\frac{1}{2}\theta}$$

$$\sum_{n=1}^{N} \cos n\theta = -\frac{1}{2} + \frac{\sin\left(N+\frac{1}{2}\right)\theta}{2\sin\frac{1}{2}\theta}$$

Derivation

Lagrange's trigonometric identities may be derived by starting with the complex sum

$$\sum_{n=0}^{N} z^n = \frac{1-z^{N+1}}{1-z}.$$

Setting $z = e^{i\theta}$ and equating the imaginary parts of both sides gives the first identity; equating the real parts of both sides gives the second.

Euler's Formula

According to Euler, we should regard the complex exponential e^{it} as related to the trigonometric functions $cos(t)$ and $sin(t)$ via the following inspired definition:

$e^{it} = cos\ t + i\ sin\ t$ where as usual in complex numbers $i^2 = -1$.

The justification of this notation is based on the formal derivative of both sides, namely

$$\frac{d}{dt}(e^{it}) = i(e^{it}) = i\ cos\ t + i^2\ sin\ t$$

$$= i\ cos\ t - sin\ t\ since\ i^2 = -1$$

$$\frac{d}{dt}(cos\ t + i\ sin\ t) = -\ sin\ t + i\ cos t \text{ since } i \text{ is a constant.}$$

along with the initial value of 1 for both sides at t = 0, assuming $e^0 = 1$ holds for complex values too.

The motivation for looking at this combination comes from the link between point in the plane with coordinates (x, y) and complex numbers formed by the relation $z = x + iy$, since z becomes the combination $r\cos\theta + ir\sin\theta$, which suggests that the combination may be interesting to look at (unit circle has r = 1).

This turns out to be a very important unification and simplification of many results in both trigonometry and calculus, in which the formula leads us to correct manipulations. As well, people use it in algebra and also in signal processing. This is illustrated first for some trig identities and then some differentiation and integration results which are otherwise hard to compute.

1. An Amusing Equation: From Euler's formula with angle π, it follows that the equation:

$$e^{i\pi} + 1 = 0$$

which involves five interesting math values in one short equation.

2. Trig Identities: The notation suggests that the following formula ought to hold:

$$e^{is} \cdot e^{it} = e^{i(s+t)}$$

which converts to the addition laws for cos and sin in components:

$$\cos(s+t) = \cos s \cos t - \sin s \sin t$$
$$\sin(s+t) = \sin s \cos t + \cos s \sin t.$$

This codifies the addition laws in trig in a way you can always recover.

We can also express the trig functions in terms of the complex exponentials e^{it}, e^{-it} since we know that cos(t) is even in t and sin(t) is odd in t. This reads as follows:

$$e^{it} = \cos t + i \sin t, \; e^{-it} = \cos t - i \sin t$$

so adding (and dividing by 2) or subtracting (and dividing by 2 i) gives:

$$\cos t = \frac{e^{it} + e^{-it}}{2}, \; \sin t = \frac{e^{it} - e^{-it}}{2i}.$$

Another manipulation suggested by the notation:

$$e^{it}e^{-it} = e^{it-it} = e^{0} = 1$$

which leads to

$$1 = (\cos t + i \sin t)(\cos(-t) + i \sin(-t)) = (\cos t + i \sin t)(\cos t - i \sin t)$$
$$= \cos^2 t - i^2 \sin^2 t = \cos^2 t + \sin^2 t.$$

There are many other uses and examples of this beautiful and useful formula. The following is known as DeMoivre's Theorem:

For any positive integer $n, e^{int} = (e^{it})^n = (\cos t + i \sin t)^n$.

This allows us to find solutions to algebra equations like $z^3 = 1$ by viewing 1 via Euler as having angle 0, 2π, 4π. Then the three solutions are found to be

$$1,\ e^{\frac{2\pi i}{3}} = \frac{-1+\sqrt{3}i}{2},\ e^{\frac{4\pi i}{3}} = \frac{-1-\sqrt{3}i}{2}$$

Note that you could find the solutions by factoring, from

$$z^3 - 1 = (z - 1)(z^2 + z + 1),$$

but factoring won't be easy for $z^7 - 1$, while Euler's formula works beautifully.

Application: Signal Processing Using the solutions to $z^n = 1$ which form n equally spaced points around the circle, approximations for periodic functions (signals) are found using the finite Fourier transform. For powers of 2 (and other composite numbers), a fast algorithm exists to compute these.

3. Calculus: The functions of the form $e^{at} \cos bt$ and $e^{at} \sin bt$ come up in applications often. To find their derivatives, we can either use the product rule or use Euler's formula

$$\begin{aligned}\left(\frac{d}{dt}\right)\left(e^{at} \cos bt + i\, e^{at} \sin bt\right) &= \left(\frac{d}{dt}\right)e^{(a+ib)t} = (a + ib)e^{(a+ib)t} \\ &= (a + ib)\left(e^{at} \cos bt + ie^{at} \sin bt\right) \\ &= \left(ae^{at} \cos bt - b\, e^{at} \sin bt\right) \\ &\quad + i\left(b\, e^{at} \cos bt + ae^{at} \sin bt\right)\end{aligned}$$

This finds both derivatives simultaneously and is especially nice for higher derivatives. Integration: Even better is the integral aspect: To integrate $e^{at} \cos bt$ and $e^{at} \sin bt$ simultaneously, integrate the complex exponential instead.

$$\begin{aligned}\int \left(e^{at} \cos bt + i\, e^{at} \sin bt\right) dt &= \int e^{(a+ib)t}\, dt = \frac{1}{a+ib} e^{(a+ib)t} + C \\ &= \frac{a-ib}{a^2+b^2}\left(e^{at} \cos bt + i\, e^{at} \sin bt\right) + C \\ &= \frac{a}{a^2+b^2}\, e^{at}\, (\cos bt + \frac{b}{a^2+b^2}\, e^{at} \sin bt) + C_1 \\ &\quad + i\left(-\frac{b}{a^2+b^2}\, e^{at} \cos bt + \frac{a}{a^2+b^2}\, e^{at} \sin bt + C_2\right).\end{aligned}$$

Another integration result is that any product of positive powers of cosine and sine can be integrated explicitly. From Euler's formula this becomes an algebra problem with an easy calculus part, as illustrated in the following example:

$$\int \cos^2 t\, dt = \int \left(\frac{e^{it} + e^{-it}}{2}\right)^2 dt = \int \left(\frac{e^{2it} + 2 + e^{-2it}}{4}\right) dt$$

Which can be done term-by-term

There is clearly nothing special about the power 2 or cosine alone, so any positive power of sine and cosine can be expanded and then integrated.

The complex logarithm Using polar coordinates and Euler's formula allows us to define the complex exponential as

$$e^{x+iy} = e^{x}\, e^{iy}$$

which can be reversed for any non-zero complex number written in polar form as $\rho e^{i\varphi}$ by inspection: $x = \ln(\rho)$, $y = \phi$ to which we can also add any integer multiplying 2π o y for another solution.

4. Differential equations: This formula really comes into its own when we need to solve differential equations with constant coefficients. Then the goal is to find the right numbers a, b so that the above functions which we just differentiated solve a given equation. For example, electrical circuits lead to differential equations that relate current, charge and voltage based on the circuit elements. Circuit elements are described by certain parameters like inductance, resistance, and capacitance. These become coefficients in the differential equation.

Example The differential equation $ay'' + by' + cy = 0$ can be solved by seeking exponential solutions with an unknown exponential factor. Substituting $y = e^{rt}$ into the equation gives a solution if the quadratic equation $ar^2 + br + c = 0$ holds. For lots of values of a, b, c, namely those where $b^2 - 4ac < 0$, the solutions are complex. Euler's formula allows us to interpret that easy algebra correctly.

De Moivre's Formula

Statement

For every real number θ and every positive integer n, we have

$$(\cos\theta + i\sin\theta)^n = \cos n\theta + i\sin n\theta\,.$$

Proof: We prove this theorem by induction, i.e. first we prove it for n=1 and then we prove that if $(\cos\theta + i\sin\theta)^n = \cos n\theta + i\sin n\theta$ holds for a particular value of n, then it holds for n+1 as well. This suffices to prove the theorem for every positive integer n. (Induction is a commonly-used method for proving mathematical results.) The case n=1 is trivial. Assume $(\cos\theta + i\sin\theta)^n = \cos n\theta + i\sin n\theta$ holds for n. Then we have

$$(\cos\theta + i\sin\theta)^{n+1} = (\cos\theta + i\sin\theta)^n(\cos\theta + i\sin\theta)$$

$$= (\cos n\theta + i\sin n\theta)\,(\cos\theta + i\sin\theta)$$

Where in the last step we used the induction hypothesis, i.e. the assumption that $(\cos\theta + i\sin\theta)^n = \cos n\theta + i\sin n\theta$ holds for n. Computing the product on the right yiedls

$$(\cos\theta + i\sin\theta)^{n+1} = (\cos n\theta\cos\theta - \sin n\theta\sin\theta) + i(\sin n\theta\cos\theta + \cos n\theta\sin\theta)$$

$$= \cos(n+1)\theta + i\sin(n+1)\theta$$

Where we used the addition formulas for sine and cosine in the last step. (Alternatively, we could have applied Theorem to the two factors on the right side of the previous equation.) Thus, $(\cos\theta + i\sin\theta)^n = \cos n\theta + i\sin n\theta$ holds for n+1 as well, whence it holds for every positive integer n.

QED

One reason De Moivre's Theorem is useful is that it allows us to compute large powers of complex numbers expressed in polar form without having to carry out every multiplication explicitly. The following examples show this.

Example: Use De Moivre's Theorem to compute $(1+i)^{12}$.

Solution: The polar form of $1 + i$ is $\sqrt{2}(\cos\pi/4 + i\sin\pi/4)$. Thus, by De Moivre's Theorem, we have

$$(1+i)^{12} = \left[\sqrt{2}(\cos\pi/4 + i\sin\pi/4)\right]^{12}$$
$$= \left(\sqrt{2}\right)^{12}(\cos\pi/4 + i\sin\pi/4)^{12}$$
$$= 2^6(\cos 3\pi + i\sin 3\pi)$$
$$= 64(\cos\pi + i\sin\pi) = 64(-1) = -64.$$

Example: Use De Moivre's Theorem to compute $\left(\sqrt{3}+i\right)^5$.

Solution: It is straightforward to show that the polar form of $\sqrt{3} + i$ is $2(\cos\pi/6 + i\sin\pi/6)$. Thus we have

$$\left(\sqrt{3}+i\right)^5 = \left[2(\cos\pi/6 + i\sin\pi/6)\right]^5$$
$$= 2^5(\cos\pi/6 + i\sin\pi/6)^5$$
$$= 32(\cos 5\pi/6 + i\sin 5\pi/6)$$
$$= 32\left(-\sqrt{3}/2 + 1/2\,i\right)$$
$$= -16\sqrt{3} + 16i.$$

A very important application of De Moivre's Theorem is computing nth roots of complex numbers, where n is a posiive integer. First we look at nth roots of 1, also known as *nth roots of unity*. First we note that 1 may be written in polar form as $1 = \cos 2\pi m + i\sin 2\pi m$ for every integer m, since cosine and sine are periodic with period 2π. Now consider the complex number

$z = \cos(2\pi m/n)n + i\sin(2\pi m/n)$, where n is an arbitrary positive integer. By De Moivre's Theorem we have

$$z^n = [\cos(2\pi m/n) + i\sin(2\pi m/n)]^n = \cos 2\pi m + i\sin 2\pi m = 1.$$

Thus we have shown that $\cos(2\pi m/n) + i\sin(2\pi m/n)$ is an nth root of unity. In fact, all the nth roots of unity are obtained this way by plugging in all integer values of m from 0 to n-1. (Every other integer m yields a root of unity identical to one of these.) Thus we now know how to find all n nth roots of unity for every positive integer n.

Example 3: Compute the three cube roots of unity.

Solution: The three cube roots of unity have the form cos $(2\pi m/3)$ + i sin $(2\pi m/3)$ for m=0, 1, or 2. Plugging in m=0 yields the root cos 0 + i sin 0 = 1. (It is easy to see that 1 is an nth root of unity for every integer n, since $1^n = 1$. Plugging in m=0 will always yield this root.) Plugging in m=1 yields the root cos $(2\pi m/3)$ + i sin $(2\pi m/3) = -1/2 - \sqrt{3}/2\, i$ and plugging in m=2 yields the root cos $(4\pi/3)$ + i sin $(4\pi/3) = -1/2 - \sqrt{3}/2$ i.

Example 4: Compute the four fourth roots of unity.

Solution: The four fourth roots of unity have the form cos $(2\pi m/4)$ + i sin $(2\pi m/4)$ for m=0, 1, 2, or 3. As usual, plugging in m=0 yields the root 1. Plugging in m=1 yields the root cos $\pi/2$ + i sin $\pi/2$ = i. Plugging in m=2 yields cos π + i sin π = -1. Finally, plugging in m=3 yields cos $3\pi/2$ + i sin $3\pi/2$ = -i.

The nth roots of unity have a nice geometric interpretation in terms of where they lie in the complex plane. They form a regular n-gon on the unit circle with one vertex at 1. The case n=6 is shown in the following figure.

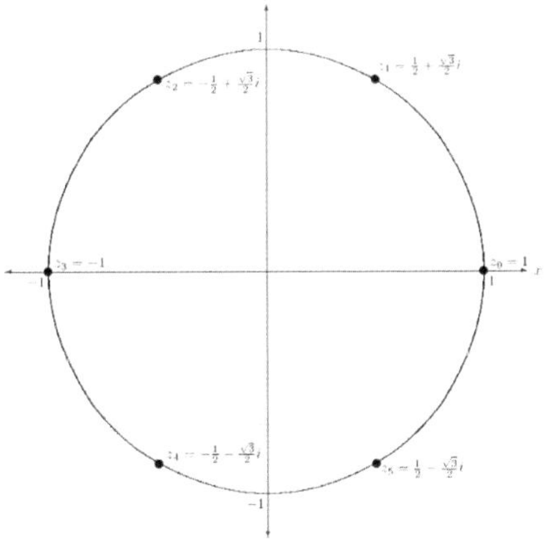

Figure: The Six Sixth Roots of Unity

So much for nth roots of unity - how about nth roots of general complex numbers? The following lemma, which is an immediate consequence of De Moivre's Theorem, tells us how to compute the n nth roots of an arbitrary complex number, given in polar form.

Lemma: Let z = r (cos θ + i sin θ) be an arbitrary complex number. Then the n nth roots w of z are each of the form

$w = \sqrt[n]{r}\ \{\cos[(\theta + 2\pi m)/n] + i \sin[(\theta + 2\pi m)/n]\}$

where m is an integer ranging from 0 to n-1.

Proof: If we apply De Moivre's Theorem to each of the alleged nth roots of z, we find

$$\begin{aligned} w^n &= \left(\sqrt[n]{r}\ \{\cos[(\theta+2\pi m)/n] + i\sin[(\theta+2\pi m)/n]\}\right)^n \\ &= r\left\{\cos[(\theta+2\pi m)/n] + i\sin[(\theta+2\pi m)/n]\right\}^n \\ &= r\left[\cos(\theta+2\pi m) + i\sin(\theta+2\pi m)\right] \\ &= r(\cos\theta + i\sin\theta) = z. \end{aligned}$$

Thus, each of these is indeed an nth root of z. Since we have found n distinct roots, this must be all of them.

QED

It should be noted that in lemma as well as our derivation of the formula for the nth roots of unity, we have implicitly used a very powerful result, known as the *Fundamental Theorem of Algebra*. This theorem says that every polynomial of degree n factors completely into a product of n linear factors over the complex numbers. Among other things, this theorem implies that every polynomial of degree n has at most n complex roots. Thus, there can be at most n nth roots of a fixed complex number a, since these roots satisfy the degree-n polynomial equation $x^n - a = 0$.

Example: Compute the two square roots of i.

Solution: It is easy to see that i has the polar form cos π/2 + i sin π/2. Its square roots are cos π/4 + i sin π/4 = $\sqrt{2}/2 + \sqrt{2}/2$ i and cos 3π/4 + i sin 3π/4 = $-\sqrt{2}/2 - \sqrt{2}/2$ i.

Example: Compute the three cube roots of -8.

Solution: Since -8 has the polar form 8 (cos π + i sin π), its three cube roots have the form $\sqrt[3]{8}$ {cos[(π + 2π m)/3] + i sin[(π + 2π m)/3]} for m=0, 1, and 2. Thus the roots are 2 (cos π/3 + i sin π/3) = 1 + √3 i, 2 (cos π + i sin π) = -2, and 2 (cos 5π/3 + i sin 5π/3) = 1 - $\sqrt{3}$ i.

Formulae for Cosine and Sine Individually

Being an equality of complex numbers, one necessarily has equality both of the real parts and of the imaginary parts of both members of the equation. If *x*, and therefore also cos *x* and sin *x*, are real numbers, then the identity of these parts can be written using binomial coefficients. This formula was given by 16th century French mathematician François Viète:

$$\sin(nx) = \sum_{k=0}^{n} \binom{n}{k} (\cos x)^k (\sin x)^{n-k} \sin\frac{(n-k)\pi}{2}$$

$$\cos(nx) = \sum_{k=0}^{n} \binom{n}{k} (\cos x)^k (\sin x)^{n-k} \cos \frac{(n-k)\pi}{2}$$

In each of these two equations, the final trigonometric function equals one or minus one or zero, thus removing half the entries in each of the sums. These equations are in fact even valid for complex values of x, because both sides are entire (that is, holomorphic on the whole complex plane) functions of x, and two such functions that coincide on the real axis necessarily coincide everywhere. Here are the concrete instances of these equations for $n = 2$ and $n = 3$:

$$\cos 2x = (\cos x)^2 + ((\cos x)^2 - 1) \qquad = 2(\cos x)^2 - 1$$

$$\sin 2x = 2(\sin x)(\cos x)$$

$$\cos 3x = (\cos x)^3 + 3\cos x((\cos x)^2 - 1) \quad = 4(\cos x)^3 - 3\cos x$$

$$\sin 3x = 3(\cos x)^2 (\sin x) - (\sin x)^3 \qquad = 3\sin x - 4(\sin x)^3.$$

The right hand side of the formula for cos nx is in fact the value $T_n(\cos x)$ of the Chebyshev polynomial T_n at cos x.

Failure for Non-integer Powers, and Generalization

De Moivre's formula does not hold for non-integer powers. The derivation of de Moivre's formula above involves a complex number raised to the integer power n. If a complex number is raised to a non-integer power, the result is multiple-valued. For example, when $n = \frac{1}{2}$, de Moivre's formula gives the following results:

for $x = 0$ the formula gives $1^{1/2} = 1$, and

for $x = 2\pi$ the formula gives $1^{1/2} = -1$.

This assigns two different values for the same expression $1^{1/2}$, so the formula is not consistent in this case.

On the other hand, the values 1 and −1 are both square roots of 1. More generally, if z and w are complex numbers, then

$$(\cos z + i\sin z)^w$$

is multi-valued while

$$\cos wz + i \sin wz$$

is not. However, it is always the case that

$$\cos wz + i \sin wz$$

is one value of

$$(\cos z + i\sin z)^w$$

Analogues in other Settings

Hyperbolic Trigonometry

Since cosh x + sinh x = e^x, an analog to de Moivre's formula also applies to the hyperbolic trigonometry. For all $n \in Z$,

$$(\cosh x + \sinh x)^n = \cosh nx + \sinh nx$$

Also, if $n \in Q$, then one value of $(\cosh x + \sinh x)^n$ will be $\cosh nx + \sinh nx$.

Quaternions

To find the roots of a quaternion there is an analogous form of de Moivre's formula. A quaternion in the form

$$d + a\hat{\mathbf{i}} + b\hat{\mathbf{j}} + c\hat{\mathbf{k}}$$

can be represented in the form

$$q = k(\cos\theta + \varepsilon\sin\theta) \qquad \text{for } 0 \le \theta < 2\pi.$$

In this representation,

$$k = \sqrt{d^2 + a^2 + b^2 + c^2}$$

and the trigonometric functions are defined as

$$\cos\theta = \frac{d}{k} \quad \text{and} \quad \sin\theta = \pm\frac{\sqrt{a^2 + b^2 + c^2}}{k}.$$

In the case that $a^2 + b^2 + c^2 \neq 0$,

$$\varepsilon = \pm\frac{a\hat{\mathbf{i}} + b\hat{\mathbf{j}} + c\hat{\mathbf{k}}}{\sqrt{a^2 + b^2 + c^2}}$$

that is, the unit vector. This leads to the variation of De Moivre's formula:

$$q^n = k^n(\cos n\theta + \varepsilon\sin n\theta).$$

Example

To find the cube roots of

$$Q = 1 + \hat{\mathbf{i}} + \hat{\mathbf{j}} + \hat{\mathbf{k}},$$

write the quaternion in the form

$$Q = 2\left(\cos\frac{\pi}{3} + \varepsilon\sin\frac{\pi}{3}\right) \qquad \text{where } \varepsilon = \frac{\hat{\mathbf{i}} + \hat{\mathbf{j}} + \hat{\mathbf{k}}}{\sqrt{3}}.$$

Then the cube roots are given by:

$$\sqrt[3]{Q} = \sqrt[3]{2}(\cos\theta + \varepsilon \sin\theta) \qquad \text{for } \theta = \frac{\pi}{9}, \frac{7\pi}{9}, \frac{13\pi}{9}.$$

2×2 matrices

Consider the following matrix $A = \begin{pmatrix} \cos\phi & \sin\phi \\ -\sin\phi & \cos\phi \end{pmatrix}$. Then $\begin{pmatrix} \cos\phi & \sin\phi \\ -\sin\phi & \cos\phi \end{pmatrix}^n = \begin{pmatrix} \cos n\phi & \sin n\phi \\ -\sin n\phi & \cos n\phi \end{pmatrix}$. This fact (although it can be proven in the very same way as for complex numbers) is a direct consequence of the fact that the space of matrices of type $\begin{pmatrix} a & b \\ -b & a \end{pmatrix}$ is isomorphic to the space of complex numbers.

Morrie's Law

Morrie's Law is the following trigonometric curiousity:

$$\cos(20^\circ)\cos(40^\circ)\cos(80^\circ) = \frac{1}{8}$$

This is notable because the angle measure 20° has radian measure $\frac{\pi}{9}$ and $\cos(\pi/9)$ isn't one of the usual constructible numbers, and yet here we have a nice, clean algebraic relationship between the numbers $\cos(20^\circ)$, $\cos(2.20^\circ), \cos(4.20^\circ)$, and $1/8$.

We will prove the more general identity: For any θ we have

$$2^n \prod_{i=0}^{n-1} \cos(2^i\theta) = \frac{\sin(2^n\theta)}{\sin(\theta)}$$

Morrie's Law is then the special case where and $\theta = \pi/9$. (Note that $\sin(8\pi/9) = \sin(\pi/9)$.)

Another special case ($n = 2$ and $\theta = \pi/5$) gives

$$4\cos(\pi/5)\cos(2\pi/5) = \frac{\sin(4\pi/5)}{\sin(\pi/5)} = 1$$

as $\sin(4\pi/5) = \sin(\pi/5)$ and thus

$$\cos(\pi/5)\cos(2\pi/5) = \frac{1}{4} \text{ or } \cos(36^\circ)\cos(72^\circ) = \frac{1}{4}.$$

The proof itself is easy: Using the well-known identity $\sin(2X) = 2\sin X \cos X$, we have

$$2^n \prod_{i=0}^{n-1} \cos(2^i\theta)$$

$$= 2\cos(\theta).2\cos(2\theta).2\cos(2^2\theta)...2\cos(2^{n-1}\theta)$$

$$= \frac{\sin(2\theta)}{\sin(\theta)}.\frac{\sin(4\theta)}{\sin(2\theta)}.\frac{\sin(8\theta)}{\sin(4\theta)}...\frac{\sin(2^n\theta)}{\sin(2^{n-1}\theta)}$$

$$= \frac{\sin(2^2\theta)}{\sin(\theta)}$$

and we're done.

References

- Mukhopadhyay, Utpal (August 2006). "Some interesting features of hyperbolic functions". Resonance. 11 (8): 81–85. doi:10.1007/BF02855783How-to-find-half-angle-identities-for-tangent, trigonometry: dummies.com, Retrieved 31 March 2018
- Lial, Margaret L.; Hornsby, John; Schneider, David I.; Callie J., Daniels (2008). College Algebra and Trigonometry (4th ed.). Boston: Pearson/Addison Wesley. p. 792. ISBN 9780321497444Lagranges-trigonometric-identities: nanopedia.davidgsimpson.com, Retrieved 11 April 2018
- Brand, Louis (October 1942). "The roots of a quaternion". The American Mathematical Monthly. 49 (8): 519–520. doi:10.2307/2302858. JSTOR 2302858Morries-law: mathnow.wordpress.com, Retrieved 12 June 2018

6

Trigonometric Equation and its Solutions

An equation which involves one or more trigonometric ratios of unknown angles is known as a trigonometric equation. The aim of this chapter is to explore trigonometric equations and the methods for finding the solutions of such equations.

Trigonometric Equation

An equation involving one or more trigonometric ratios of unknown angles is called a trigonometric equation. A trigonometric equation can be written as Q_1 (sin θ, cos θ, tan θ, cot θ, sec θ, cosec θ) = Q_2 (sin θ, cos θ, tan θ, cot θ, sec θ, cosec θ), where Q_1 and Q_2 are rational functions.

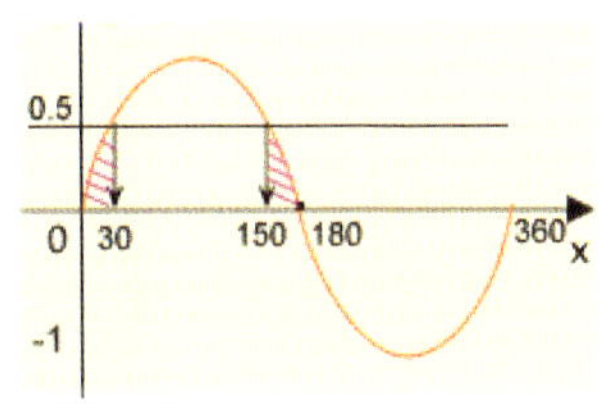

Example: Let us consider an equation $\cos^2 x - 4 \sin x = 1$.

This is a trigonometric equation and not an identity as it is not satisfied for all values of x e.g. the equation is not satisfied at $(2n + 1)\pi/4$.

A trigonometric equation that holds true for any angle is called a trigonometric identity. There are other equations, though, that only are true for certain angles. They are generally known as conditional equations, We'll learn some techniques for solving general equations, as well as how to derive an infinite number of solutions to an equation based on a single solution to that equation.

Only a few simple trigonometric equations can be easily solved without a calculator. Often one might encounter an equation like tan(x) = 3.2. Such an equation has no simple answer that can be memorized. It would be tedious to use a calculator and try numerous values for x until you found one that gave a solution close to 3.2. For problems like these, the inverse trigonometric functions are helpful. The inverse trigonometric functions are the same as the trigonometric functions, except x and y are reversed. For example, another way to say sin(y) = x is y = arcsin (x). The arcsine relation is not a function, though, because it assigns more than one element of the range to each element of the domain.

For example, $sin(y) = \frac{1}{2}$ has solutions of y = 30 degrees, 150 degrees, 390 degrees, and so on. When the range is restricted, however, then arcsine is a function, and is written with a capital letter, Arcsine. Using the inverse trigonometric functions, it becomes possible (with a calculator) to solve nearly any trigonometric equation without difficulty.

Solution of a Trigonometric Equation

All possible values of unknown which satisfy the given equation are called solution of the given equation.

For complete solution "all possible values" satisfying the equation must be obtained.

When we try to solve a trigonometric equation, we try to find out all sets of values of θ, which satisfy the given equation. Sometimes, in simple equations and when it is easy to draw a graph of an equation, one can find out the solution simply by viewing the graph.

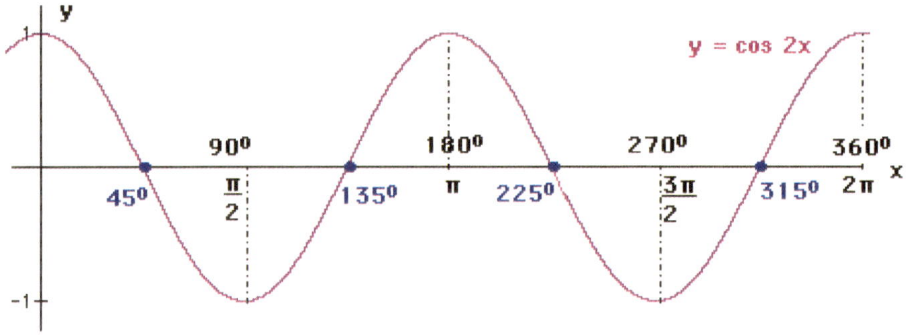

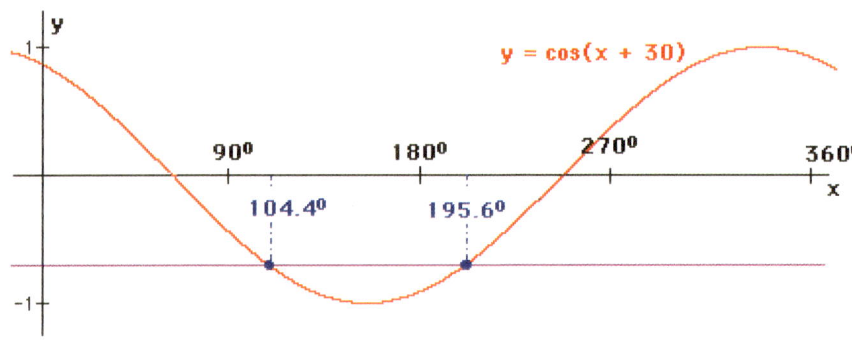

Period of a Function

A function f(x) is said to be periodic if there exists $T > 0$ such that $f(x+T) = f(x)$ for all x in the domain of definition of f(x). If T is the smallest positive real number such that $f(x+T) = f(x)$, then it is called the period of f(x).

The trigonometric functions such as sin, cos and tan are periodic functions.

Illustration: We try to search for the solutions of the equation $\sin\theta = 0$ other than $\theta = 0$. By seeing the equation, one might straight away reach at the conclusion that $\theta = 0$ is the only solution. But, in case of trigonometric equations, it is important to rule out all possibilities so as to reach at the correct solution.

Let OX be the initial line

Let $\angle POX = \theta$ and $OP = r$

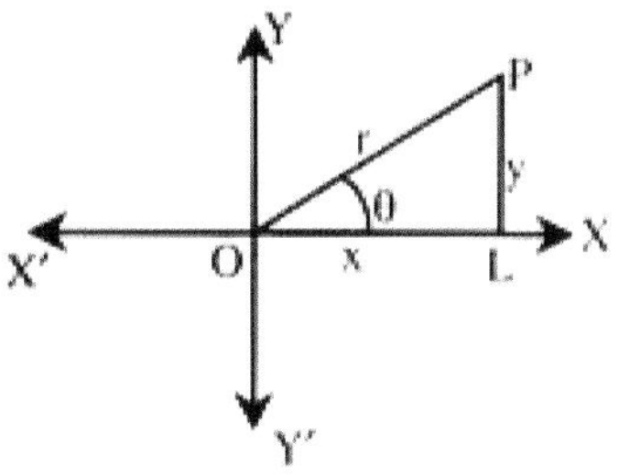

From ΔPOL,

sin θ = PL/OP = y/r.

Now sin θ = 0

⇒ y/r = 0; ⇒ y = 0.

This is possible only when OP coincides with OX or OX'.

When OP coincides with OX, θ = 0, ± 2π, ± 4π and ± 6π

And when OP coincides with OX', θ = ± π, ± 3π, ± 5π

Thus from above it follows that at sin θ = 0

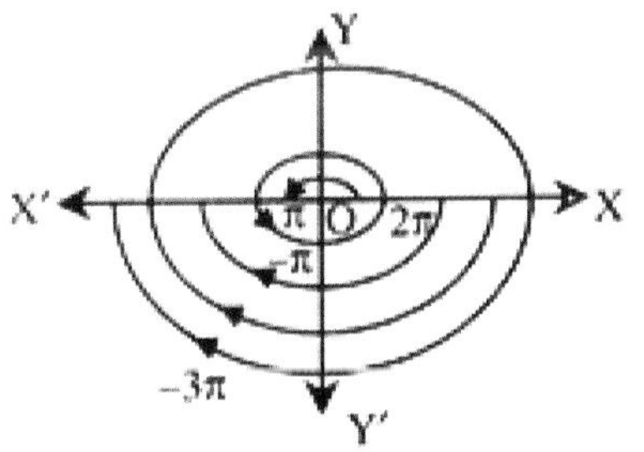

θ = nπ, where n = 0, ±1, ±2,.........

We call θ = nπ, a general solution of the trigonometric equation sin θ = 0, because for all values of n, this solution satisfies the given equation.

Illustration: General solution of cos θ = 0

cos θ = 0 ⇒ x = π/2.

This is possible only when OP coincides with OY or OY'

When OP coincides with OY,

θ = π/2, 5π/2, 9π/2 or, -3π/2, -7π/2

when OP coincides with OY'

θ = -3π/2, -7π/2 or, -π/2, -5 /2

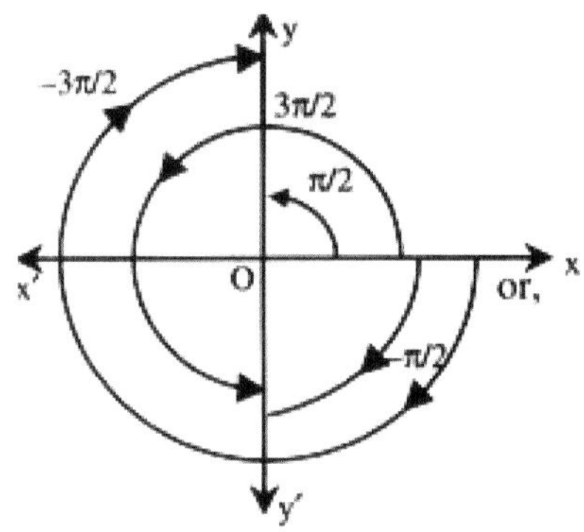

Thus from above if follows that the general solution of cos θ = 0 is θ (2n+1) π/2, where n = 0, ±1, ±2 ....

Basic trigonometric equations have the form

$$\sin x = a,\ \cos x = a,\ \tan x = a,\ \cot x = a,$$

where x is an unknown, a is any real number.

Equation $\sin x = a$

If $|a| > 1$, the equation $\sin x = a$ has no solutions.

If $|a| > 1$, the general solution of the equation $\sin x = a$ is written as $x = (-1)^n \arcsin a + \pi n, n \in \mathbb{Z}$.

This formula contains two branches of solutions:

$$x_1 = \arcsin a + 2\pi n,\ x_2 = \pi - \arcsin a + 2\pi n,\ n \in \mathbb{Z}.$$

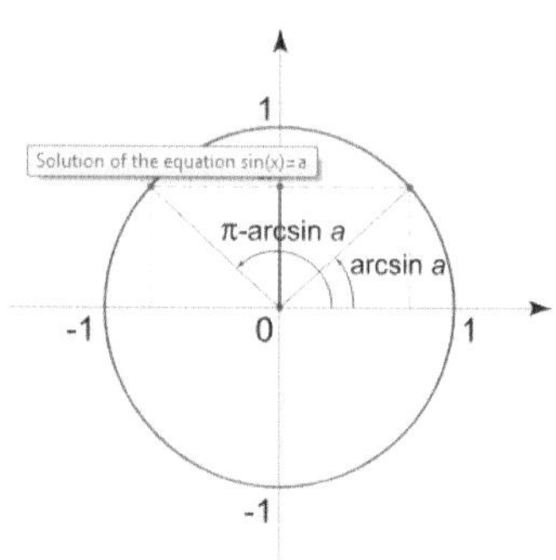

In the simple case $\sin x = 1$ the solution has the form $x = \pi / 2 + 2\pi n, n \in \mathbb{Z}$.

Similarly, the solution of the equation $\sin x = -1$ is given by $x = -\pi / 2 + 2\pi n, n \in \mathbb{Z}$.

Case $\sin x = 0$ (zeroes of the sine)

$$x = \pi n, n \in \mathbb{Z}.$$

Equation $\cos x = a$

If $|a|>1$, the equation $\cos x = a$ has no solutions.

If $|a|>1$, the general solution of the equation $\cos x = a$ has the form

$$x = \pm\arccos a + 2\pi n, n \in \mathbb{Z}.$$

This formula includes two sets of solutions:

$$x_1 = \arccos a + 2\pi n, x_2 = -\arccos a + 2\pi n, n \in \mathbb{Z}.$$

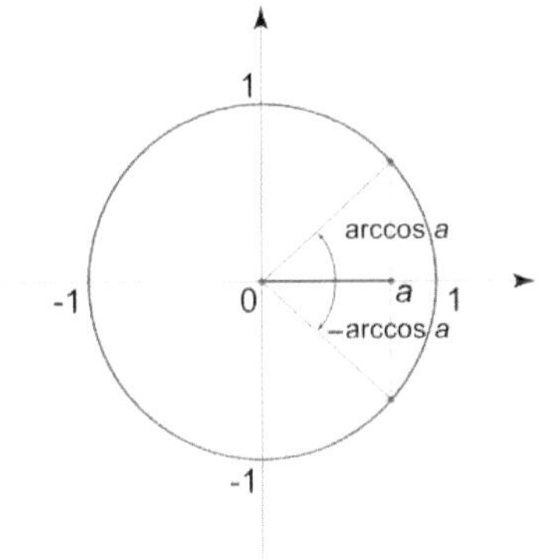

In the case $\cos x = 1$, the solution is written as

$$x = 2\pi n, n \in \mathbb{Z}.$$

Case $\cos x = -1$

$$x = \pi + 2\pi n, n \in \mathbb{Z}.$$

Case $\cos x = 0$ (zeroes of the cosine)

$$x = \pi / 2 + \pi n, n \in \mathbb{Z}.$$

Equation $\tan x = a$

For any value of a, the general solution of the equation $\tan x = a$ has the form

$$x = \arctan a + \pi n, n \in \mathbb{Z}.$$

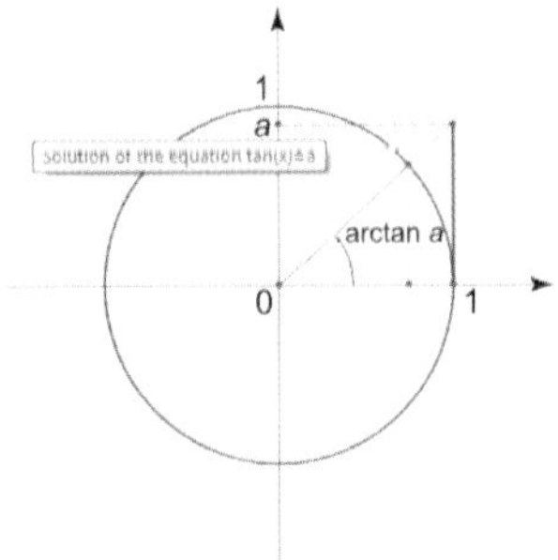

Case $\tan x = 0$ (zeroes of the tangent)

$x = \pi n, n \in \mathbb{Z}.$

Equation $\cot x = 0$

For any value of a, the general solution of the trigonometric equation $\cot x = 0$ is written as

$x = \text{arccot}\, a + \pi n, n \in \mathbb{Z}.$

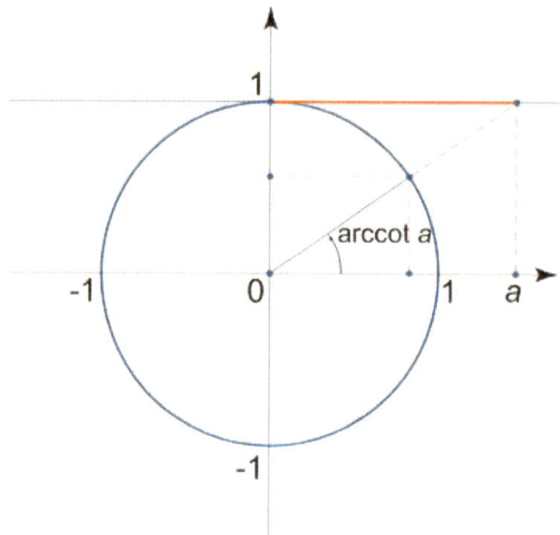

Case $\cot x = 0$ (zeroes of the cotangent)

$x = \pi / 2 + \pi n, n \in \mathbb{Z}.$

General Solution of the Equation sin θ = k.

We know that when sin θ = k, k has to be such that $-1 \leq k \leq 1$

We can always find some $\alpha \in [-\pi/2, \pi/2]$

As sin (- π)/2 = -1 & sin π /2 = 1, such that sin θ = k, i.e. α = sin-1k

i.e. sin θ = sin α, α ∈ [– π /2, π /2]

⇒ sin θ – sin α = 0

⇒ 2 sin {(θ – α)/2} cos {θ + α)/2} = 0

from the above equation to be satisfied, either sin {(θ – α)/2) = 0

and consequently ((θ – α)/2) = integral multiple of π

∴ θ – α = 2nπ

i.e. θ = 2nπ + α

θ = 2nπ + (–1)2n α where n = 0, ±1, ±2

or, cos $\{(\theta + \alpha)/2\} = 0$

i.e. $\{(\theta + \alpha)/2\}$ = any odd multiple of $\pi/2$

i.e. $\{(\theta + \alpha)/2\} = (2n + 1)\pi/2$

i.e. $\theta = (2n + 1)\pi - \alpha$

$\Rightarrow \theta = (2n + 1)\pi + (-1)2n+1\,\alpha$

From ($\theta = 2n\pi + (-1)2n\,\alpha$ where $n = 0, \pm1, \pm2$) and ($\theta = (2n + 1)\pi + (-1)2n+1\,\alpha$), we conclude that

$\theta = n\pi + (-1)n\,\alpha$, where n is integral multiple, is the general solution of the equation $\sin\theta = k$

Trigonometric Equations with their general Solutions:

Trigonometrical equation	General Solution
$\sin\theta = 0$	Then $\theta = n\pi$
$\cos\theta = 0$	$\theta = (n\pi + \pi/2)$
$\tan\theta = 0$	$\theta = n\pi$
$\sin\theta = 1$	$\theta = (2n\pi + \pi/2) = (4n+1)\pi/2$
$\cos\theta = 1$	$\theta = 2n\pi$
$\sin\theta = \sin\alpha$	$\theta = n\pi + (-1)n\alpha$, where $\alpha \in [-\pi/2, \pi/2]$
$\cos\theta = \cos\alpha$	$\theta = 2n\pi \pm \alpha$, where $\alpha \in (0, \pi]$
$\tan\theta = \tan\alpha$	$\theta = n\pi + \alpha$, where $\alpha \in (-\pi/2, \pi/2]$
$\sin 2\theta = \sin 2\alpha$	$\theta = n\pi \pm \alpha$
$\cos 2\theta = \cos 2\alpha$	$\theta = n\pi \pm \alpha$
$\tan 2\theta = \tan 2\alpha$	$\theta = n\pi \pm \alpha$

If α is assumed to be the least positive value of θ which satisfies two given trigonometrical equations, then the general value of θ will be $2n\pi + \alpha$.

Illustration: Find the general solution of equation $\sin\theta = 1/2$

Solution: We know that $\sin\theta = 1/2 = \sin\pi/6$.

So, general solution of the given equation is $\theta = n\pi + (-1)n\pi/6$, $n \in 0, \pm1, \pm2$

Illustration: Solve the equation $\sin 6x + \sin 4x = 0$.

Solution: Applying the formulae for the sum of sine, i.e.

$\sin A + \sin B = \sin (A+B)/2 \,.\, \cos (A-B)/2$, we have

$$\sin 5x \cos x = 0$$

If 'x' is a solution for the equation, then at least one of the following equations is true:

$$\sin 5x = 0 \text{ or } \cos x = 0$$

Conversely, if x is a solution of one of the equation $\sin 5x = 0$ or $\cos x = 0$ then it is solution of equation $\sin 5x \cos x = 0$ as well. Thus equation $\sin 5x = 0$ or $\cos x = 0$ is equivalent to the equation $\sin 5x = 0$ or $\cos x = 0$. The solutions of the equation $\sin 5x = 0$ or $\cos x = 0$ are given by

$x = n\pi/5$, $x = (2n+1)\pi/2$, where $n = 0, \pm 1, \pm 2$ ......

All these values of x and only these values are the solutions of the original equation.

Illustration: Find the most general value of θ where $\sin \theta = -\sqrt{3}/2$ and $\tan \theta = \sqrt{3}$.

Solution: Signs assume great importance in case of trigonometric functions. Students generally tend to mention the general solutions of sin and $\tan \theta$, which is wrong as that would not give us the entire solution.

$\sin \theta$ is negative in 3rd and 4th Quadrant and $\tan \theta$ is positive in 1st and 3rd Quadrant.

So common is 3rd Quadrant and at $\theta = 4\pi/3$ both are satisfied.

$\therefore$ The general solution is $2n\pi + 4\pi/3$.

This is because in interval $[0, 2\pi]$ it is satisfied only at $4\pi/3$. Again in $[2\pi, 4\pi]$ it is satisfied at $2\pi + 4\pi/3$ and so on.

Hence, the general solution of the equation is $2n\pi + 4\pi/3$.

Illustration: Find general solution of $\cos 3\theta = \sin 2\theta$.

Solution: This can be solved by 2 different ways.

Method 1: We can write the given equation as

$\cos 3\theta = \cos (\pi/2 - 2\theta)$

$\Rightarrow 3\theta = 2n\pi + (\pi/2 - 2\theta)$, where $n = 0, \pm 1, \pm 2$ ......

or $5\theta = 2n\pi + \pi/2$ and also $\theta = 2n\pi - \pi/2$

or $\theta = (4n + 1)\pi/10$ and

$\theta = (4n-1)\pi/2$, where $n \in I$ ...... (A)

Method 2: $\sin 2\theta = \sin (\pi/2 - 3\theta)$

$2\theta = n\pi + (-1)^n (\pi/2 - 3\theta)$.

Case I: When n is even, $n = 2m$, where $m = 0, \pm 1, \pm 2$ ......

$2\theta = 2m\pi + \pi/2 - 3\theta$

$\theta = (4m+1)\pi/10$, where $m \in I$ .... (B)

Case II: When n is odd, $n = (2m + 1)$

$2\theta = (2m+1)\pi - (\pi/2 - 3\theta)$

$\theta = -(4m + 1)\pi/2$, where $m = 0, \pm1, \pm2$ ...... (B)

Note: No doubt solutions obtained by both methods for odd values of n are different but as shown in the chart below, you can see that all possible values of θ are obtainable by both the given solutions:

From B	From A
for $m = 0, \theta = -\pi/2$,	for $n = 0, \theta = -\pi/2$
for $m = 1, \theta = -5\pi/2$,	for $n = 1, \theta = +3\pi/2$
for $m = 2, \theta = -9\pi/2$,	for $n = 2, \theta = +7\pi/2$
for $m = -1, \theta = -3\pi/2$,	for $m = -1, \theta = -5\pi/2$
for $m = -2, \theta = 7\pi/2$	for $m = -2, \theta = -9\pi/2$

General solution of sin2 θ = k, where k ∈ [0, 1]

Given that $\sin^2\theta = k$, $k \in [0, 1]$

We can find some α such that

$\Rightarrow \sin^2\theta = \sin^2\alpha$ where $\alpha = \sin^{-1}\sqrt{k}$

i.e. $(\sin\theta - \sin\alpha)(\sin\theta + \sin\alpha) = 0$

either $\sin\theta - \sin\alpha = 0$

$$\theta = n\pi + (-1)^n \alpha, \text{ where } n = 0, \pm1, \pm2$$

or, $\sin\theta + \sin\alpha = 0$

$\sin\theta = -\sin\alpha$

$$\theta = n\pi - (-1)^n \alpha, \text{ where } n = 0, \pm1, \pm2$$

From $\theta = n\pi + (-1)^n \alpha$, where $n = 0, \pm1, \pm2$ and $\theta - n\pi - (-1)^n \alpha$, where $n = 0, \pm1, \pm2$ we get the general solution of equation the given

$\theta = n\pi \pm \alpha$ where $n = 0, \pm1, \pm2$ ...... and $\alpha = \sin^{-1}\sqrt{k}$

Illustration: Solve the equation $7\tan^2\theta - 9 = 3\sec^2\theta$

Solution: Given, $7\tan^2\theta - 9 = 3\sec^2\theta$

or, $7\tan^2\theta - 9 = 3(1 + \tan^2\theta)$

or, $4\tan^2\theta = 12$

or, $\tan^2\theta = 3$

or, $\tan^2\theta = (\tan \pi/3)^2$

$\Rightarrow \theta = n\pi + \pi/3$, where $n = 0, \pm1, \pm2$ ....

Note: We cannot define a unique method of solving trigonometric equations. In each case, success in solving a trigonometric equation depends, in particular, on the knowledge and application ability of trigonometric formula and the practice of solving problems. Many trigonometric formulas are true equalities for all the values of the variable's appearing in them.

Illustration: Solve the equation: $\cos \theta = 0$

Solution: We can solve it to get two forms

$\cos \theta = 0 \Rightarrow \theta = (2n + 1)\pi/2$

$\cos \theta = \cos \pi/2 \Rightarrow \theta = 2n\pi + \pi/2$

or, $\theta = (4n + 1)\pi/2$.

Important: Following tips and steps will help you systematically solve trigonometric equations.

- Try to reduce equation in terms of one single trigonometric ratios preferably $\sin \theta$ or $\cos \theta$. If we have choice to convert a problem in sine or cosine then cosine form is convenient compared to sine form. This is because in general solution of sine, we will have to deal with $(-1)^n$ which is inconvenient compared to the dealing of + obtained in cosine form.
- Factorize the polynomial in terms of these ratios.
- For LHS to be zero solve for each factor. And write down general solution, for each factor based on the standard results that are derived earlier in this section.

e.g. $\sin \theta - k_1 = 0 \Rightarrow \theta = n\pi + (-1)^n \sin^{-1} k_1$

$\cos \theta - k_2 = 0 \Rightarrow \theta = 2n\pi + \cos^{-1} k_2$.

Caution: You must check that $k_1, k_2 \in [-1, 1]$. Don't blindly write it as it is since it will be absurd if they do not belong to $[-1, 1]$.

Illustration: Solve the equation $5\sin \theta - 2 \cos^2\theta - 1 = 0$

Solution: Given, $5 \sin \theta - 2 \cos^2\theta - 1 = 0$

or, $5 \sin \theta - 2 (1 - \sin^2\theta) - 1 = 0$

or, $2 \sin^2 \theta + 5 \sin \theta - 3 = 0$

or, $(\sin \theta + 3) (2 \sin \theta - 1) = 0$

$\therefore \sin \theta = -3$ or $\sin \theta = 1/2$

First consider the case when $\sin\theta = -3$.

But this case is not possible as the range of sine is [-1, 1].

When $\sin\theta = 1/2$

Then we have $\sin\theta = \sin\pi/6$.

$\Rightarrow \theta = n\pi + (-1)^n \pi/6$ where $n = 0, \pm1, \pm2$ ....

Note: Never divide by any expression which is zero. e.g. If the given equation is $(\sin\theta - \cos\theta)$ (A) = (B) where A and B denote trigonometric equations then you can divide by $(\sin\theta - \cos\theta)$ only when $\theta \neq n\pi + \pi/4$

Illustration: Solve the equation $\tan\theta + \sec\theta = \sqrt{3}$

Solution: $\tan\theta + \sec\theta = \sqrt{3}$

Then $(\sin\theta)/(\cos\theta) + 1/(\cos\theta) = \sqrt{3}$

or, $\cos(\theta + \pi/6) = \cos\pi/3$

General solution $\theta + \pi/6 = 2n\pi \pm \pi/3, n \in I$

Taking positive sign, $\theta + \pi/6 = 2n\pi + \pi/3$

$\Rightarrow \theta = 2n\pi + \pi/6$

Taking negative sign, $\theta + \pi/6 = 2n\pi - \pi/3$

$\Rightarrow \theta = 2n\pi - \pi/2$

i.e. $\theta = (4n - 1)\pi/2$.

But the solution obtained is correct only if, $\cos\theta \neq 0$ otherwise $((\sin\theta)/(\cos\theta) + 1/(\cos\theta) = \sqrt{3})$ is not defined.

i.e. $\theta \neq$ odd multiple of $\pi/2$

$\Rightarrow \theta \neq (4n - 1)\pi/2$.

Hence the general solution will be $\theta = 2n\pi + \pi/6$ only where $n = 0, \pm1, \pm2$ ......

Note: Domain of an equation should not change. In case it changes, necessary changes must be made in the general solution.

Illustration: Solve the equation $\tan 5\theta = \tan 3\theta$

Solution: Now $\tan 5\theta = \tan 3\theta$

$\Rightarrow 5\theta = n\pi + 3\theta$

or, $2\theta = n\pi$

$\theta = n\pi/2$, where $n = 0, \pm 1, \pm 2$

putting $n = 0$ gives $\theta = 0$, original equation is satisfied

putting $n = 1$ we obtain $\theta = \pi/2$ equation becomes $\tan 5\pi/2 = \tan 3\pi/2$.

Equation is not defined for odd multiple of $\pi/2$.

Hence we conclude that $\theta = 2n\pi$, where $n = 0, \pm 1, \pm 2$

7

Uses and Applications of Trigonometry

Trigonometry has applications in the fields of astronomy, architecture, engineering, geodesy, physical sciences, seismology, etc. The diverse uses and applications of trigonometry in different areas have been thoroughly discussed in this chapter. It also includes topics like application of right triangle, application of radian measure, application of trigonometric graphs, etc.

Application of Right Triangle

Many problems involve right triangles. We often need to use the trigonometric ratios to solve such problems.

Example - Finding the Height

Find h for the given triangle.

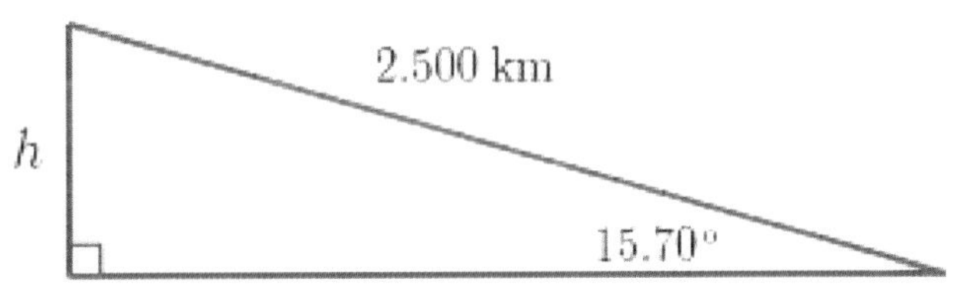

$$\sin 15.70^o = \frac{h}{2.500}$$

$$h = 2.500 \sin 15.70^o$$
$$= 0.6765 \text{ km}$$

Example - Solving Triangles

Solve the triangle ABC, given that $A = 35^\circ$ and $c = 15.67$.

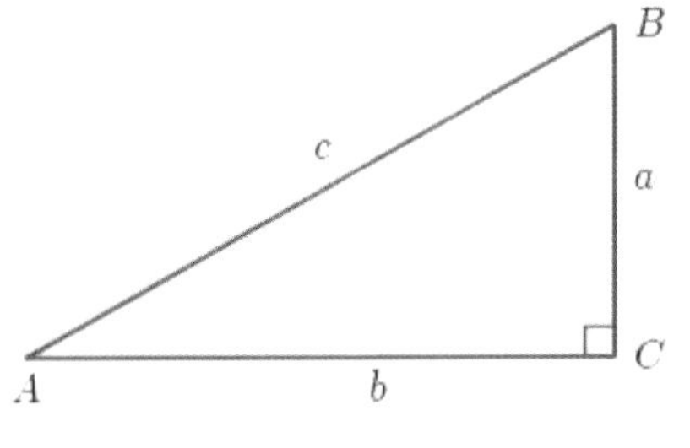

To "solve" a triangle means to find the unknown sides and angles.

We have, for the values given:

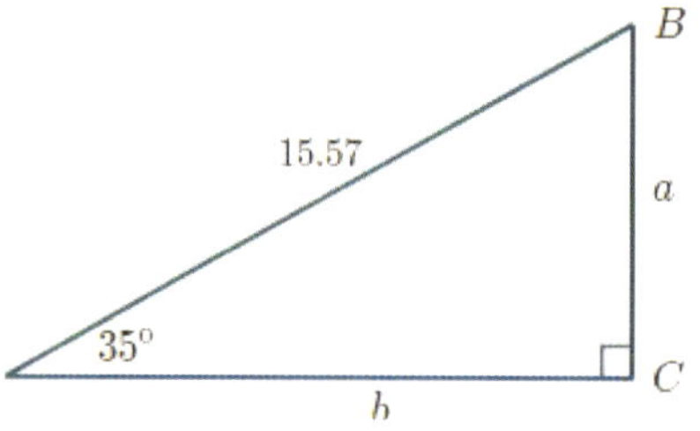

In this example, we need to find side lengths a and b and angle B. Note $C = 90°$.

$$\sin 35^{o} = \frac{a}{c} = \frac{a}{15.67}$$

So $a = 15.67 \sin 35^{o} = 8.99$

$$\cos 35^{o} = \frac{b}{c} = \frac{b}{15.67}$$

So $b = 15.67 \cos 35^{o} = 12.84$

Angle $B = 90^{o} - 35^{o} = 55^{o}$.

So $a = 8.99, b = 12.84$ and $B = 55^{o}$. We have found all the unknowns.

Angles of Elevation and Depression

In surveying, the angle of elevation is the angle from the horizontal looking up to some object:

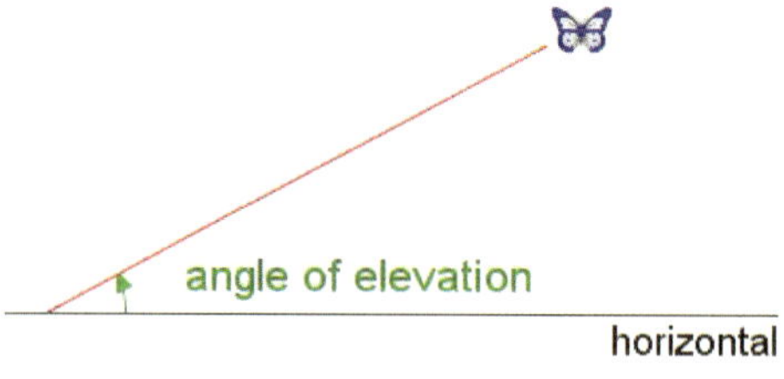

The angle of depression is the angle from the horizontal looking down to some object:

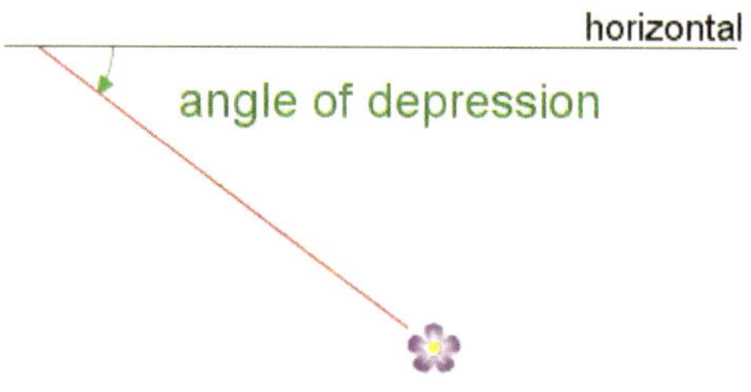

Example -

The angle of elevation of an aero plane is 23°. If the airplane's altitude is 2500 m, how far away is it?

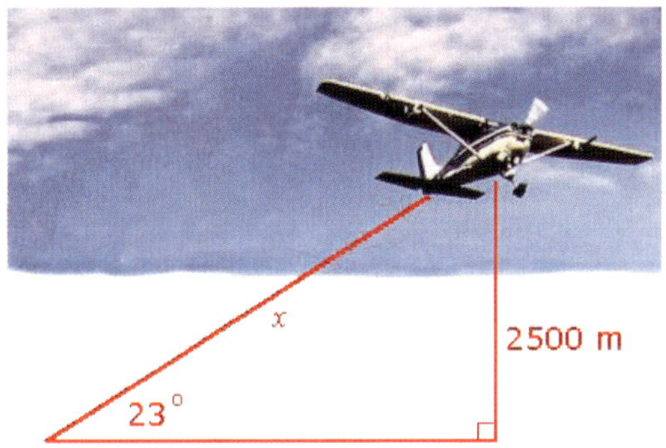

Let the distance be x. Then $\sin 23^{o} = \dfrac{2500}{x}$

$$x = \frac{2500}{\sin 23^{o}} = 6400 \text{ m}$$

Example -

You can walk across the Sydney Harbour Bridge and take a photo of the Opera House from about the same height as top of the highest sail.

This photo was taken from a point about 500 m horizontally from the Opera House and we observe the waterline below the highest sail as having an angle of depression of 8°. How high above sea level is the highest sail of the Opera House?

This is a simple tan ratio problem.

$$\tan 8^{o} = \frac{h}{500}$$

So

$$h = 500 \tan 8^{\circ} = 70.27 \text{ m}.$$

So the height of the tallest point is around 70 m.

[The actual height is 67.4 m.]

Application of Radian Measure

We will see some of the common applications of radian measure, including arc length, area of a sector of a circle, and angular velocity.

Arc Length

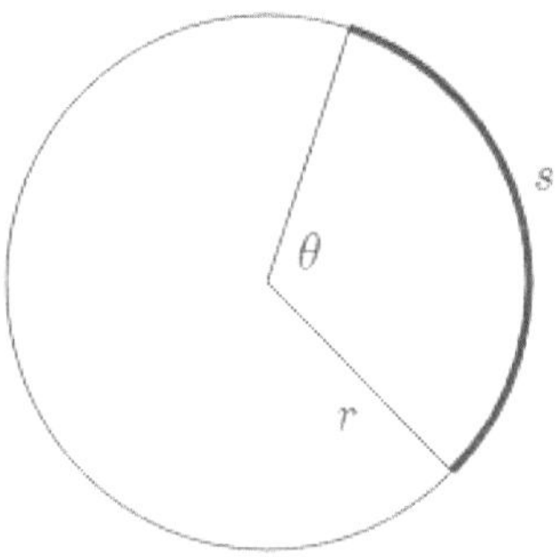

The length, s, of an arc of a circle radius r subtended by θ (in radians) is given by:

$s = r\,\theta$

If r is in meters, s will also be in meters. Likewise, if r is in cm, s will also be in cm.

Example:

Find the length of the arc of a circle with radius 4 cm and central angle 5.1 radians.

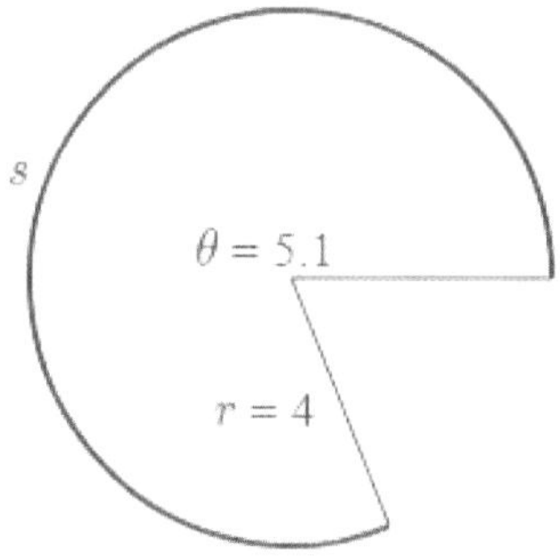

We need to find length s.

$s = r\theta$

$= 4 \times 5.1$

$= 20.4$ cm

Area of a Sector

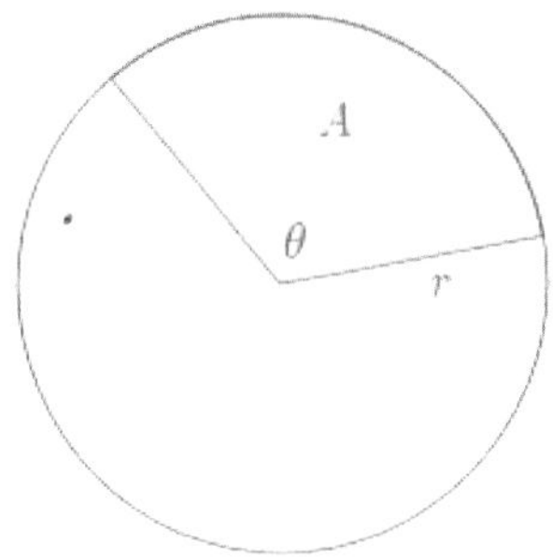

Area, *A*, of a sector of a circle.

The area of a sector with central angle θ (in radians) is given by:

$$\text{Area} = \frac{\theta r^2}{2}$$

If r is measured in cm, the area will be in cm^2. If r is in m, the area will be in m^2

Example:

Find the area of the sector with radius 7 cm and central angle 2.5 radians.

This is the area we need to find:

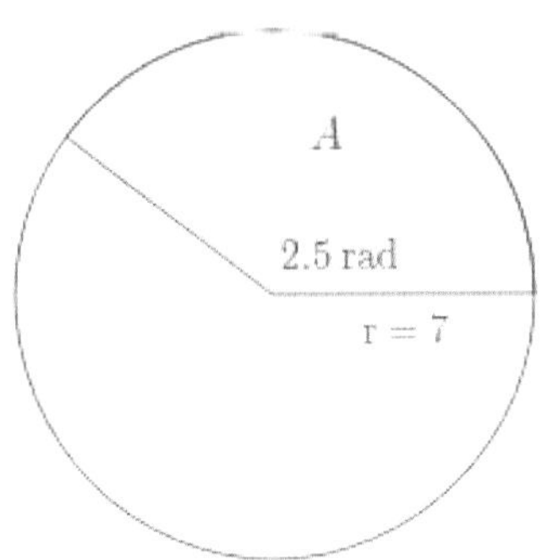

Area, a sector of a circle, radius 7 and central angle 2.5 rad.

$$\text{Area} = \frac{\theta r^2}{2} = \frac{2.5 \times 7^2}{2} = 61.25\,\text{cm}^2$$

Angular Velocity

The time rate of change of angle θ by a rotating body is the angular velocity, written ω (omega). It is measured in radians/second.

If v is the linear velocity (in m/s) and r is the radius of the circle (in m), then

$v = r\omega$

If r is in cm, v will be in cm/s.

Example:

A bicycle with tyres 90 cm in diameter is travelling at 25 km/h. What is the angular velocity of the tyre in radians per second?

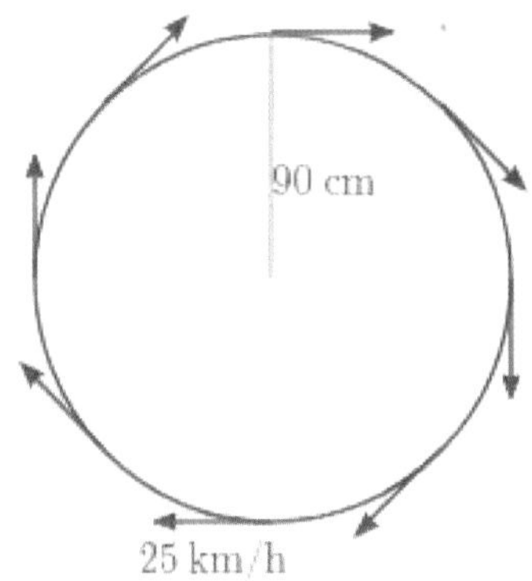

Bicycle wheel, radius 90 cm. The arrows represent the linear speed of 25 km/h

We learned that linear velocity for a wheel rotating at ω rad/s is given by:

$v = r\,\omega$

The units are a mix of cm and km. Let's present everything in meters.

We need to convert v to m/s first.

$$\begin{aligned} 25\text{ km/h} &= 25000\text{ m/h} \\ &= \frac{25000}{3600}\text{ m/s} \\ &= 6.94444\text{ m/s} \end{aligned}$$

Also, we have

$$r = \frac{90\text{ cm}}{2} = 45\text{ cm} = 0.45\text{ m}$$

$$\text{So } w = \frac{v}{r} = \frac{6.94444}{0.45} = 15.43\text{ rad/s}$$

Interactive Linear Velocity Applet

Background

A car is going around a circular track of radius 0.5 km. The speedometer in the car shows the (magnitude) of the linear velocity.

At first, the car goes around the track once in just over 6 minutes. It's angular velocity is 1 rad/min or one complete revolution in $2\pi = 6.28$ min.

The distance travelled in this time is the circumference of the circle, $C = 2\pi r = 2\pi(0.5) = 3.14$ km.

So the car is travelling at $\dfrac{3.14\,\text{km}}{6.28\,\text{min}} = 0.5\,\text{km/min} = 30\,\text{km/h}$.

The linear velocity showing on the speedo is 30 km/h.

Exercises:

1. A section of side walk is a circular sector of radius 1.25 m and central angle 50.6°. What is the area of this section of sidewalk?

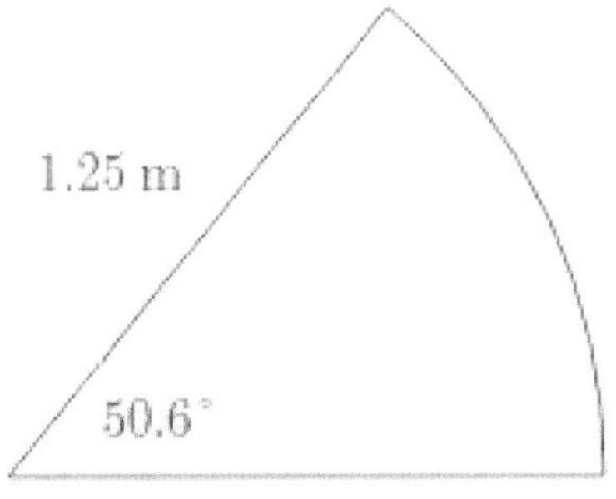

Circular sector, radius 1.25 m, central angle 56°.

First we must convert 50.6° to radians:

$$50.6^\circ = 50.6 \times \frac{\pi}{180} = 0.8831 \text{ radians}$$

$$\begin{aligned}\text{Area} &= \frac{\theta r^2}{2}\\ &= \frac{0.8831 \times 1.25^2}{2}\\ &= 0.690\,\text{m}^2\end{aligned}$$

2. A cam is in the shape of a circular sector with radius 1.875 cm and central angle 165.58°. What is the perimeter of the cam?

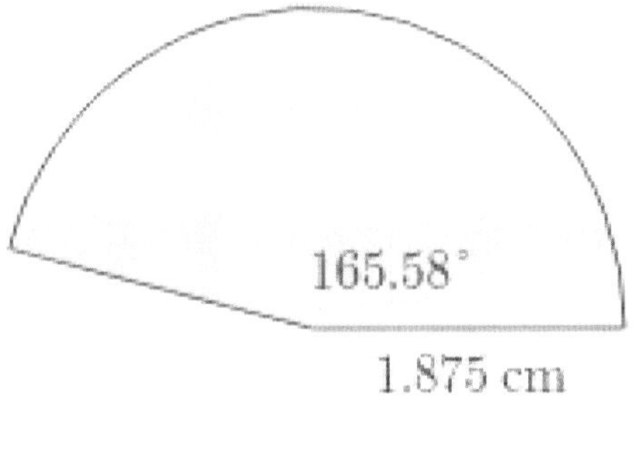

Circular sector, radius 1.875 cm, central angle 165.58°.

The length of the arc is given by $s = r\theta$.

First we must convert 165.58° into radians:

$$165.58^{\circ} = 165.58 \times \frac{\pi}{180} = 2.8899 \text{ radians}.$$

$$\text{So arc length is}: s = 1.875 \times 2.8899 = 5.419 \text{ cm}.$$

$$\text{So the perimeter of the cam is}:$$

$$2 \times 1.875 + 5.419 = 9.169 \text{ cm}.$$

3. The roller on a computer printer make 2200 rev/min. What is its angular velocity?

Angular velocity is:

$$w = 2200 \frac{r}{\min} \times \frac{2\pi}{60} = 230.4 \text{ rad/s}.$$

4. The propeller on a motorboat is rotating at 130 rad/s. What is the linear velocity of a point on the tip of a blade if the blade is 22.5 cm long?

Linear velocity = $v = \omega r$

In this example, $\omega = 130$ rad/s and $r = 0.225$ m

So the linear velocity is:

$v = 130 \times 0.225 = 29.3 \text{ ms}^{-1}$.

Note 1: ms^{-1} is an equivalent way of writing m/s. This comes from the index laws where the rule is $s^{-1} = \frac{1}{s}$.

Note 2: It is common in physics to write velocity using ms^{-1} and the units for acceleration as ms^{-2}.

5. The sweep second hand of a watch is 15.0 mm long. What is the linear velocity of the tip?

The second hand rotates 2π every minute, so per second we have:

$$w = \frac{2\pi}{60} = \frac{2\pi}{30} \text{rad/s}$$

$$\text{and } r = 0.015 \text{m}.$$

So

$$\begin{aligned} v &= wr \\ &= \frac{\pi}{30}(0.015) \\ &= 0.00157 \text{m/s}. \end{aligned}$$

Application of Trigonometric Graphs

Simple Harmonic Motion

Any object moving with constant angular velocity or moving up and down with a regular motion can be described in terms of simple harmonic motion.

The displacement, d, of an object moving with SHM, is given by:

$$d = R \sin \omega t$$

Where R is the radius of the rotating object and ω is the angular velocity of the object.

NOTE: We may need to use one the following, depending on the situation:

$$d = R \cos \omega t \quad d = R \sin(\omega t + \alpha) \quad d = R \cos(\omega t + \alpha)$$

Example:

A point on a cam is 8.30 cm from the centre of rotation. Sketch 2 cycles of d as a function of t, given that d = 0 cm when t = 0 s and ω = 3.20 rad/s.

Since $d = R \sin \omega t$, we have

$d = 8.30 \sin 3.20t$.

This sine curve will have amplitude 8.30 and period given by

$$\text{Period} = \frac{2\pi}{b} = \frac{2\pi}{3.20} = 1.96$$

So the sketch will be:

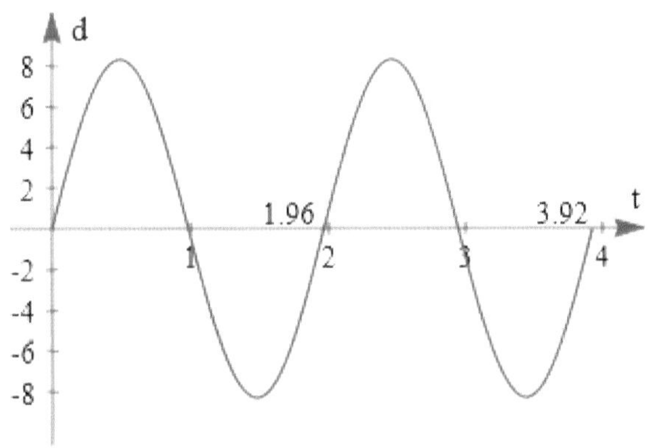

Graph of $d = 8.30 \sin(3.20t)$

Example 2

The voltage of an alternating current circuit is given by

$e = E \cos(\omega t + \alpha)$.

Sketch 2 cycles of the voltage as a function of time if

$$E = 80\,V, w = 377\,\text{rad/s and } \alpha = \frac{\pi}{2}.$$

We need to sketch:

$e = E\cos(\omega t + \alpha)$

$e = 80\cos(377t + \pi/2)$

This will have amplitude = 80,

$$\text{Period} = \frac{2\pi}{377} = 0.0167 \text{ and}$$

$$\text{phase shift} = -\frac{c}{b} = -\frac{\frac{\pi}{2}}{377} = -0.00417$$

So our sketch is:

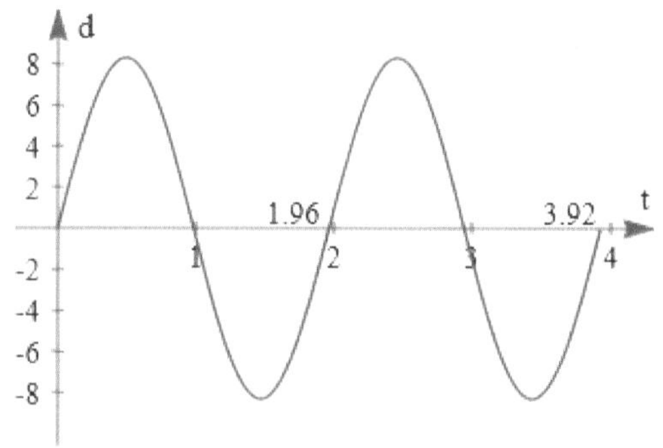

Graph of $e = 80\cos(377t + \pi/2)$

Example 3

The signal received by a radio is given by

$e = 0.014\cos(2\pi ft)$,

Where e is in volts and f is in Hz.

Draw 2 cycles of e for f=950 kHz.

Now, f=950000.

We need to find the wavelength to be able to draw 2 cycles.

We have: $b = 2\pi f$ and we know that

$$\text{Period} = \frac{2\pi}{b} = \frac{2\pi}{2\pi \times 950000} = 1.053 \times 10^{-6}$$

So 2 wavelengths will be: $2\times 1.053\times 10^{-6} = 2.105\times 10^{-6}$

So, graphing for $0 < t < 2.105\times 10^{-6}$, we have:

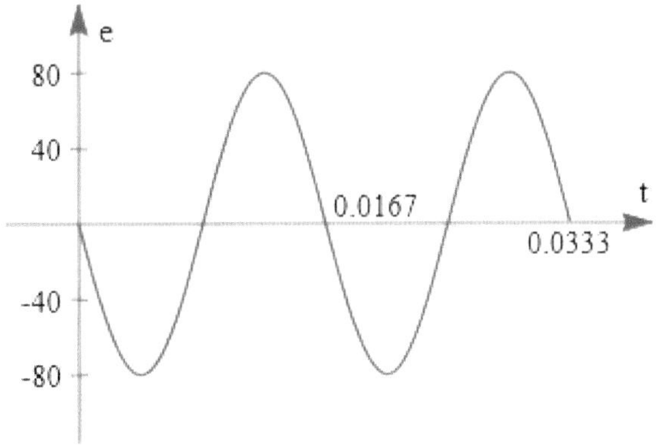

Graph of e = 0.014 cos(2π ×950000t). Units for t are 10^{-6} s

Angular Velocity

Another important result in this section is:

The angular velocity ω (in radians per second) of a rotating object is given by:

$\omega = 2\pi f$

where f is the frequency of the motion, in cycles per second.

Exercises

1. A satellite is orbiting the earth so that its displacement D north of the equator is given by

$$D = A\sin(\omega t + \alpha).$$

Sketch 2 cycles of D as a function of t if

$$A = 500\text{ km}, \omega = 3.60\, rad/hr \text{ and} \alpha = 0.$$

So we need to sketch: D = 500 sin(3.6t)

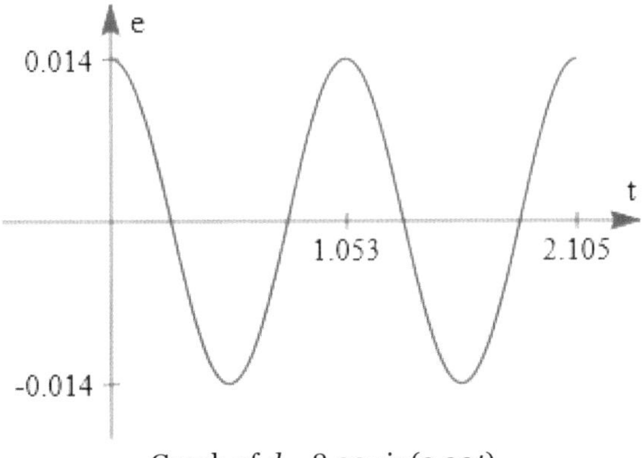

Graph of d = 8.30 sin(3.20t)

Note: Be careful with units! If frequency is in cycles per minute, then the angular velocity will be in radians per minute. Also, be careful with "kilo".

2. Using $e = E\cos(\omega t + \alpha)$, sketch 2 cycles of the voltage as a function of time if

$$E = 170 \text{ V}, w = 120\pi \text{ rad/s and } \alpha = -\frac{\pi}{3}.$$

$$\text{So } e = 170\cos(120\pi t - \frac{\pi}{3})$$

Amplitude is 170 V.

$$\text{Period} = \frac{2\pi}{b} = \frac{2\pi}{120\pi} = \frac{1}{60} = 0.016667$$

$$\text{Phase shift} = -\frac{c}{b} = -\frac{-\frac{\pi}{3}}{120\pi} = \frac{1}{360} = 0.0027778$$

So our cosine curve will be shifted to the right by approximately 0.0028 seconds.

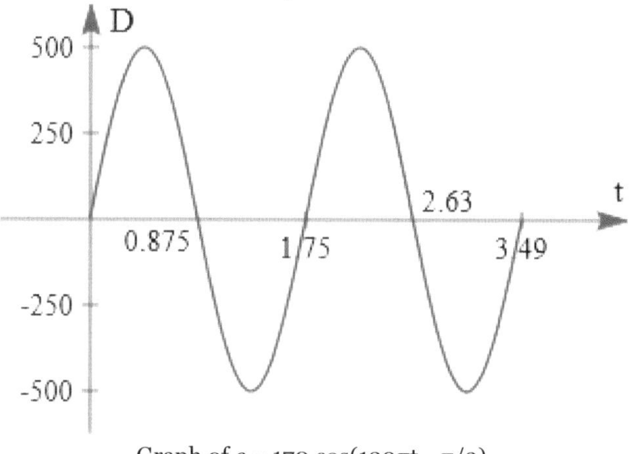

Graph of $e = 170\cos(120\pi t - \pi/3)$

Uses of Trigonometry

Trigonometry may not have its direct applications in solving practical issues, but it is used in various things that we enjoy so much. For example music, as you know sound travels in waves and this pattern though not as regular as a sine or cosine function, is still useful in developing computer music. A computer cannot obviously listen to and comprehend music as we do, so computers represent it mathematically by its constituent sound waves. And this means sound engineers need to know at least the basics of trigonometry. And the good music that these sound engineers produce is used to calm us from our hectic, stress full life.

Trigonometry can be used to Measure the Height of a Building or Mountains

If you know the distance from where you observe the building and the angle of elevation you can easily find the height of the building. Similarly, if you have the value of one side and the angle of depression from the top of the building you can find and another side in the triangle, all you need to know is one side and angle of the triangle.

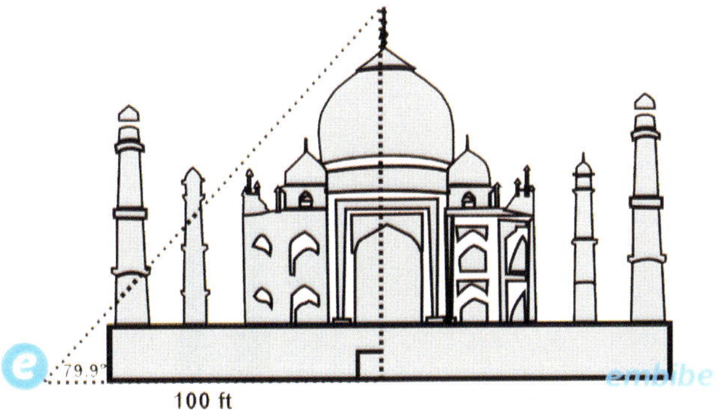

Example

The distance from where the building is observed is 90ft90ft from its base and the angle of elevation to the top of the building is ∘35∘. Now find the height of the building.

Solution:

Given:

- Distance from the building is 90feet from the building.
- The angle of elevation from to the top of the building is 35∘.

To solve and find the height of the tower by recalling the trigonometric formulas. Here, the angle and the adjacent side length are provided. So, using the formula of tan.

$$\tan 35^{0} = \frac{Opposite\ side}{Adjacent\ side}$$

$$\tan 35^{0} = \frac{h}{90}$$

$$h = 90 \times \tan 35^{0}$$

$$h = 90 \times 0.4738$$

$$h = 42.64\,feet$$

Thus, the height of the building is 42.64feet.

Trigonometry in video games

Have you ever played the game, Mario? When you see him so smoothly glide over the road blocks. He doesn't really jump straight along the Y axis, it is a slightly curved path or a parabolic path that he takes to tackle the obstacles on his way. Trigonometry helps Mario jump over these obstacles.

As you know Gaming industry is all about IT and computers and hence Trigonometry is of equal importance for these engineers.

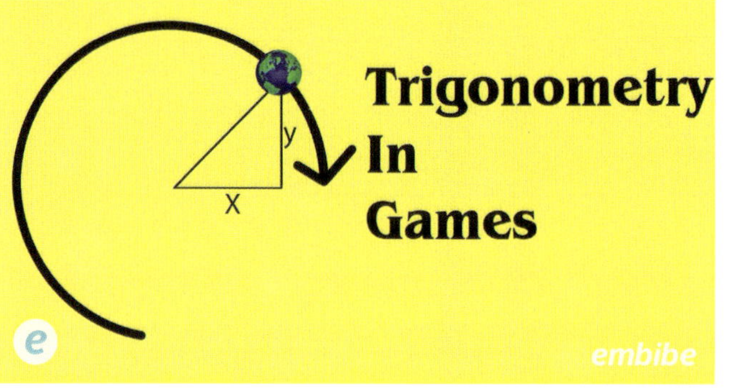

Trigonometry in construction

In construction we need trigonometry to calculate the following:

- Measuring fields, lots and areas;
- Making walls parallel and perpendicular;
- Installing ceramic tiles;
- Roof inclination;
- The height of the building, the width length etc. and the many other such things where it becomes necessary to use trigonometry.

Architects use trigonometry to calculate structural load, roof slopes, ground surfaces and many other aspects, including sun shading and light angles.

Trigonometry in Physics

In physics, trigonometry is used to find the components of vectors, model the mechanics of waves (both physical and electromagnetic) and oscillations, sum the strength of fields, and use dot and cross products. Even in projectile motion you have a lot of application of trigonometry.

Trigonometry in Marine Engineering

In marine engineering trigonometry is used to build and navigate marine vessels. To be more specific trigonometry is used to design the Marine ramp, which is a sloping surface to connect lower and higher level areas, it can be a slope or even a staircase depending on its application.

- Trigonometry in Astronomy: Measuring distances to objects within our Galaxy is not always a straightforward task – we cannot simply stretch out a measuring tape between two objects and read off the distance. Instead, a number of techniques have been developed that enable us to measure distances to stars without needing to leave the Solar System. One such method is trigonometric parallax, which depends on the apparent motion of nearby stars compared to more distant stars, using observations made six months apart.
- Trigonometry in Architecture: Mathematics makes the design of buildings safer and more accurate. Trigonometry is especially important in architecture because it allows the architect to calculate distances and forces related to diagonal elements. Of the six functions in basic trigonometry, the sine, cosine and tangent are the most important to architecture because they allow the architect to easily find the opposite and adjacent values related to an angle or hypotenuse, translating a diagonal vector into horizontal and vertical vectors.
- Trigonometry in Chemistry: Chemists use trigonometric functions to accurately describe the angles that are created when atoms bond together to form molecules. chemists can represent the shape of the molecule most accurately in a three-dimensional model, which can be obtained using trigonometric functions. The molecule is represented within a triangular prism so that chemists can work out the equations that determine the interior angles and length of the triangle's three sides, which are called the opposite, adjacent and hypotenuse.
- Trigonometry in Meteorology: Meteorology is the scientific study of the atmosphere. A meteorologist uses trigonometry for analysis, modelling and prediction. Weather balloons are used to help measure temperature, humidity and wind. These balloons can typically ascend to 100,000 feet and drift for 125 miles. They carry a device called a radiosonde, which takes humidity, pressure and temperature measurements. The initial height of the balloon can be calculated from the tangent function with the use of a ground crew's distance and angle of sight. The tangent function is also used to calculate the balloon's height when it's between two observation stations. The balloon's height, angle, distance and travel time can be used in calculations to help estimate wind speed.
- Trigonometry in Engineering: Trigonometry is not just a subject to be studied in a classroom with no real world practical applications. Engineers of various types use the fundamentals of trigonometry to build structures/systems, design bridges and solve scientific problems. When all of the measurements of the structure are known the engineer can begin building and defining the scope of the project they are undertaking.
- Trigonometry in Carpentry: Carpentry calls for trigonometry more than you might think. Every time a carpenter makes an angled cut, the measurement of the angle or the adjoining lines must be figured out. Trigonometry is used in many other carpentry applications, including site layout tasks that require using angular measurements. These tasks might include laying out building foundation lines and determining elevations by completing trigonometric levelling.
- Trigonometry in Biology: Trigonometry comes up in many aspects of biology. One example

is X-ray crystallography, a technique used to determine the three dimensional structure of molecules. X-ray crystallography has been used to determine the atomic structure of thousands of biologically important molecules including vitamins, proteins, and perhaps most famously DNA. X-ray crystallography exploits the fact that when an x-ray is passed through a crystal, it is diffracted according the crystal's atomic structure.

- Trigonometry in Forensics: Forensics use trigonometry to analyze crime scenes and collect clues. They first check for any bullet holes in ceilings, floors and furniture. Next They try and re-enact the crime scene in order to try and presume what happened. this is where trigonometry comes in. Trigonometry helps them to suggest where the suspect or victim was located at the time of the incident. For example, a bullet fired into a wall, at an angle of 90° can show the investigators the size of the bullet. This can also help to determine the specific gun used by the shape of the hole in the wall. If the angle is off by even the slightest degree that makes the hole an ellipse. With a hole of this type, investigators can not only determine the size of the bullet, but, by using basic trigonometry, they can tell where the bullet came from.

Thomas Paine's Statement

The scientific principles that man employs to obtain the foreknowledge of an eclipse, or of any thing else relating to the motion of the heavenly bodies, are contained chiefly in that part of science that is called trigonometry, or the properties of a triangle, which, when applied to the study of the heavenly bodies, is called astronomy; when applied to direct the course of a ship on the ocean, it is called navigation; when applied to the construction of figures drawn by a ruler and compass, it is called geometry; when applied to the construction of plans of edifices, it is called architecture; when applied to the measurement of any portion of the surface of the earth, it is called land-surveying. In fine, it is the soul of science. It is an eternal truth: it contains the mathematical demonstration of which man speaks, and the extent of its uses are unknown.

Some Modern uses

Scientific fields that make use of trigonometry include:

> acoustics, architecture, astronomy, cartography, civil engineering, geophysics, crystallography, electrical engineering, electronics, land surveying and geodesy, many physical sciences, mechanical engineering, machining, medical imaging, number theory, oceanography, optics, pharmacology, probability theory, seismology, statistics, and visual perception

That these fields involve trigonometry does not mean knowledge of trigonometry is needed in order to learn anything about them. It *does* mean that *some* things in these fields cannot be understood without trigonometry. For example, a professor of music may perhaps know nothing of mathematics, but would probably know that Pythagoras was the earliest known contributor to the mathematical theory of music.

In *some* of the fields of endeavor listed above it is easy to imagine how trigonometry could be used. For example, in navigation and land surveying, the occasions for the use of trigonometry are in at least some cases simple enough that they can be described in a beginning trigonometry textbook. In the case of music theory, the application of trigonometry is related to work begun by Pythagoras, who observed that the sounds made by plucking two strings of different lengths are consonant if both lengths are small integer multiples of a common length. The resemblance between the shape of a vibrating string and the graph of the sine function is no mere coincidence. In oceanography, the resemblance between the shapes of some waves and the graph of the sine function is also not coincidental. In some other fields, among them climatology, biology, and economics, there are seasonal periodicities. The study of these often involves the periodic nature of the sine and cosine function.

Fourier Series

Many fields make use of trigonometry in more advanced ways than can be discussed in a single article. Often those involve what are called Fourier series, after the 18th- and 19th-century French mathematician and physicist Joseph Fourier. Fourier series have a surprisingly diverse array of applications in many scientific fields, in particular in all of the phenomena involving seasonal periodicities mentioned above, and in wave motion, and hence in the study of radiation, of acoustics, of seismology, of modulation of radio waves in electronics, and of electric power engineering.

A Fourier series is a sum of this form:

$$\square + \underbrace{\square\cos\theta + \square\sin\theta}_{1} + \underbrace{\square\cos(2\theta) + \square\sin(2\theta)}_{2} + \underbrace{\square\cos(3\theta) + \square\sin(3\theta)}_{3} + \cdots$$

where each of the squares (□) is a different number, and one is adding infinitely many terms. Fourier used these for studying heat flow and diffusion (diffusion is the process whereby, when you drop a sugar cube into a gallon of water, the sugar gradually spreads through the water, or a pollutant spreads through the air, or any dissolved substance spreads through any fluid).

Fourier series are also applicable to subjects whose connection with wave motion is far from obvious. One ubiquitous example is digital compression whereby images, audio and video data are compressed into a much smaller size which makes their transmission feasible over telephone, internet and broadcast networks. Another example, mentioned above, is diffusion. Among others are: the geometry of numbers, isoperimetric problems, recurrence of random walks, quadratic reciprocity, the central limit theorem, Heisenberg's inequality.

Fourier Transforms

A more abstract concept than Fourier series is the idea of Fourier transform. Fourier transforms involve integrals rather than sums, and are used in a similarly diverse array of scientific fields. Many natural laws are expressed by relating *rates of change* of quantities to the quantities themselves. For example: The rate of change of population is sometimes jointly proportional to (1) the present population and (2) the amount by which the present population falls short of the carrying capacity. This kind of relationship is called a differential equation. If, given this information, one tries to express population as a function of time, one is trying to "solve" the differential equation. Fourier transforms may be used to convert some differential equations to algebraic equations for which methods of solving them are known. Fourier transforms have many uses. In almost any scientific context in which the words spectrum, harmonic, or resonance are encountered, Fourier transforms or Fourier series are nearby.

Statistics, Including Mathematical Psychology

Intelligence quotients are sometimes held to be distributed according to the bell-shaped curve. About 40% of the area under the curve is in the interval from 100 to 120; correspondingly, about 40% of the population scores between 100 and 120 on IQ tests. Nearly 9% of the area under the curve is in the interval from 120 to 140; correspondingly, about 9% of the population scores between 120 and 140 on IQ tests, etc. Similarly many other things are distributed according to the "bell-shaped curve", including measurement errors in many physical measurements. Why the ubiquity of the "bell-shaped curve"? There is a theoretical reason for this, and it involves Fourier transforms and hence trigonometric functions. That is one of a variety of applications of Fourier transforms to statistics.

Trigonometric functions are also applied when statisticians study seasonal periodicities, which are often represented by Fourier series.

Number Theory

There is a hint of a connection between trigonometry and number theory. Loosely speaking, one could say that number theory deals with qualitative properties rather than quantitative properties of numbers.

$$\frac{1}{42}, \quad \frac{2}{42}, \quad \frac{3}{42}, \quad \ldots\ldots, \quad \frac{39}{42}, \quad \frac{40}{42}, \quad \frac{41}{42}.$$

Discard the ones that are not in lowest terms; keep only those that are in lowest terms:

$$\frac{1}{42}, \quad \frac{5}{42}, \quad \frac{11}{42}, \quad \ldots, \quad \frac{31}{42}, \quad \frac{37}{42}, \quad \frac{41}{42}.$$

Then bring in trigonometry:

$$\cos\left(2\pi\cdot\frac{1}{42}\right)+\cos\left(2\pi\cdot\frac{5}{42}\right)+\cdots+\cos\left(2\pi\cdot\frac{37}{42}\right)+\cos\left(2\pi\cdot\frac{41}{42}\right)$$

The value of the sum is −1, because 42 has an *odd* number of prime factors and none of them is repeated: 42 = 2 × 3 × 7. (If there had been an *even* number of non-repeated factors then the sum would have been 1; if there had been any repeated prime factors (e.g., 60 = 2 × 2 × 3 × 5) then the sum would have been 0; the sum is the Möbius function evaluated at 42.) This hints at the possibility of applying Fourier analysis to number theory.

Solving Non-trigonometric Equations

Various types of equations can be solved using trigonometry.

For example, a linear difference equation or linear differential equation with constant coefficients has solutions expressed in terms of the eigenvalues of its characteristic equation; if some of the eigenvalues are complex, the complex terms can be replaced by trigonometric functions of real terms, showing that the dynamic variable exhibits oscillations.

Similarly, cubic equations with three real solutions have an algebraic solution that is unhelpful in that it contains cube roots of complex numbers; again an alternative solution exists in terms of trigonometric functions of real terms.

Trigonometry in Criminology

Criminology is the study of the causes and consequences of criminal behaviors. When a bullet is shot out of a pistol and makes contact with its target, it leaves a mark at a certain angle. That angle and the size of the specific bullet and gun can be used to trace the projectiles trajectory to estimate the location of the shooter. Other than homicide detectives and forensics agents, this knowledge is also useful for snipers who take shots over long distances.

Bloodstain Pattern Analysis

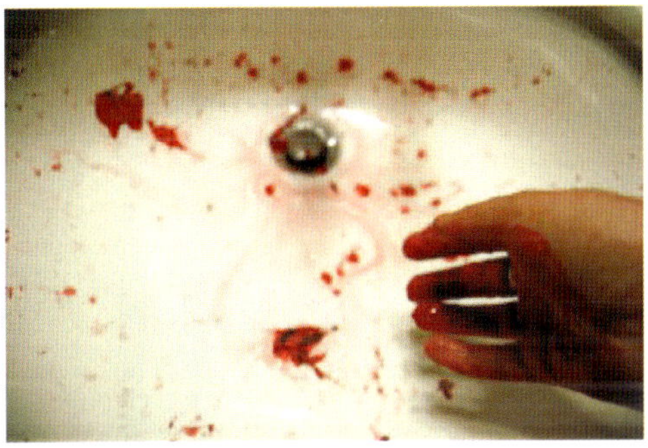

Bloodstain pattern analysis (BPA) is another tool used in forensic science that involves the study and analysis of bloodstains at a crime scene in order to recreate the actions that caused the bloodshed. The science of BPA applies scientific knowledge from other fields such as biology, physics and mathematics.

Many criminal investigations involve a death resulting from a gunshot, knife wound, etc, and it is crucial to the investigation to know the exact position of the victim or the murderer. The main objective of a BPA is to give answers to questions such as where the blood came from, how the victim and the perpetrator were positioned, how many potential perpetrators were present and whether the victim was moved or the scene changed in any way.

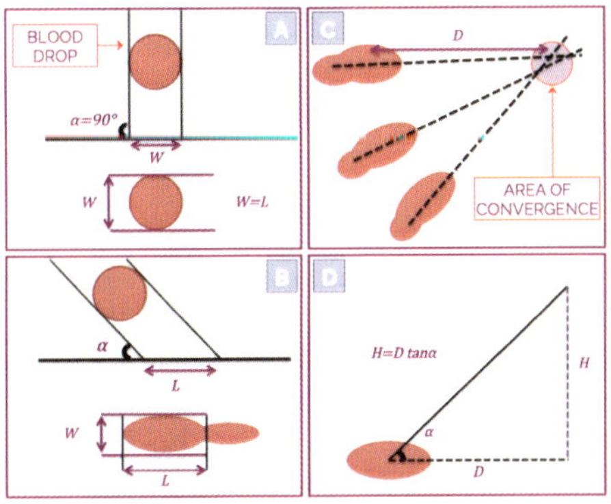

Blood pattern analysis

In the figure above, we show the general procedure used by forensic scientists to make a blood pattern analysis: in case *A*, the drop of blood hits the surface at an angle of $\alpha=90°$, giving a shape that is a circle of diameter W; and in case *B*, the blood drop strikes the floor with $\alpha\neq90°$, giving a shape that is a teardrop made up of an ellipse (diameter W and length of splatter L) and an extension resulting from the droplets' momentum. For each droplet, the forensic scientist measures W and L to determine the angle of impact α ($sin\,\alpha = W / L$), then extends the long axis of the ellipses, to find the *area of convergence*: the zone where all the lines intersect. Once they locate this area, they measure the distance D between an individual bloodstain and the area of convergence. Finally, using simple trigonometry they can determine the point of origin H of the blood spatter ($H=D \tan \alpha$). The value of H must then coincide for all the drops of blood.

Trigonometry in Marine Biology

Marine biology is the study of species inhabiting the oceans and other saline aquatic environments, and the way in which these organisms interact within their habitats. The individual career paths of marine biologists are often quite diverse; some marine biologists may work at aquariums while others manage fish hatcheries, train dolphins, study whale migration patterns, or research the effects of pollution on marine species' populations.

Marine biologists often use algebra and trigonometry to establish measurements. For example, in investigating how the light level at different depths affects the ability of algae to photosynthesize, marine biologists may utilize mathematical models, that range from logarithmic functions to hyperbolic tangent functions. Similarly, marine biologists may use geometry or trigonometry to determine the size of wild animals from a distance. Additionally, a sound knowledge of measurement equivalents is a must, as marine biologists frequently convert between standard and metric units of measure.

Trigonometry in Navigation

Trigonometry is fundamental to navigation. Trigonometry is used extensively in calculations referring to Cartesian coordinates. Cartesian coordinates are used to represent North, South, East and West directions. Through the use of Trigonometry the distance between objects, the required direction and a vessel's or aircraft's bearings can be calculated.

Compasses and Direction

A compass is a navigational instrument that shows four direction points. These points are North, South, East and West. A compass rose, showing these directions is drawn on the compass with 'N' signifying North. Compasses are essential in Navigation for it determines bearings and which way to travel. Before the compass, the position and direction at sea was done by the sighting of landmarks and celestial bodies (stars).

The invention of the compass enabled a heading to be determined when bad weather occurred in addition to latitude and longitude. Without the compass people could not accurately navigate away from land and therefore not engage in trade with other nations.

Bearings

In marine navigation, a bearing is the direction one object is from another object, usually, the direction of an object from one's own vessel. In aircraft navigation, a bearing is the actual compass direction of the forward course of the aircraft. In land navigation, a bearing is the angle between a line connecting two points. Bearings is usually measured in degrees.

Types of bearings include:

- True bearings/ Three-figure bearings
- Magnetic bearings
- Compass bearings
- Grid bearings

Trigonometry in Action

A plane leaves an airport and flies 600 km on a bearing of 145°

a) How far south of the airport is the plane?

Let x = distance south and missing angle in SAP = 35°

$\text{Cos } 35^\circ = x$

600

$x = 600 \cos 35^\circ$

$x = 491$ Km

Therefore the plane is 491 km south of the airport

b) What is the three-figure bearing of the airport from the plane?

A compass rose is drawn at the position of the plane.

NPA = 35° (alternate)

Therefore the bearing of the airport from the plane is 360°- 35° = 325°

Navigation without Trigonometry

Without Trigonometry there would be no bearings meaning that Navigation by sea, air and land will be different. However, the fact that Trigonometry is used in this field does not mean that the knowledge of Trigonometry is needed to learn about navigation. Some aspects of Navigation are easy to understand without Trigonometry such as directions, while other fields such as the music theory needed Trigonometry in order to fully understand how it works.

References

- 4-right-triangle-applications, trigonometric-functions: intmath.com, Retrieved 22 June 2018
- 5-applications-trigonometric-graphs, trigonometric-graphs: Retrieved 12 April 2018
- Applications-of-trigonometry: byjus.com, Retrieved 19 April 2018
- Real-life-applications-of-trigonometry: embibe.com, Retrieved 14 July 2018
- The-importance-of-trigonometry-in-our-lives: binarytranslator.com, Retrieved 16 March 2018
- Mathematics-and-crime-science: chalkdustmagazine.com, Retrieved 29 June 2018
- Type-math-used-marine-biology-6785: education.seattlepi.com, Retrieved 14 July 2018

Permissions

All chapters in this book are published with permission under the Creative Commons Attribution Share Alike License or equivalent. Every chapter published in this book has been scrutinized by our experts. Their significance has been extensively debated. The topics covered herein carry significant information for a comprehensive understanding. They may even be implemented as practical applications or may be referred to as a beginning point for further studies.

We would like to thank the editorial team for lending their expertise to make the book truly unique. They have played a crucial role in the development of this book. Without their invaluable contributions this book wouldn't have been possible. They have made vital efforts to compile up to date information on the varied aspects of this subject to make this book a valuable addition to the collection of many professionals and students.

This book was conceptualized with the vision of imparting up-to-date and integrated information in this field. To ensure the same, a matchless editorial board was set up. Every individual on the board went through rigorous rounds of assessment to prove their worth. After which they invested a large part of their time researching and compiling the most relevant data for our readers.

The editorial board has been involved in producing this book since its inception. They have spent rigorous hours researching and exploring the diverse topics which have resulted in the successful publishing of this book. They have passed on their knowledge of decades through this book. To expedite this challenging task, the publisher supported the team at every step. A small team of assistant editors was also appointed to further simplify the editing procedure and attain best results for the readers.

Apart from the editorial board, the designing team has also invested a significant amount of their time in understanding the subject and creating the most relevant covers. They scrutinized every image to scout for the most suitable representation of the subject and create an appropriate cover for the book.

The publishing team has been an ardent support to the editorial, designing and production team. Their endless efforts to recruit the best for this project, has resulted in the accomplishment of this book. They are a veteran in the field of academics and their pool of knowledge is as vast as their experience in printing. Their expertise and guidance has proved useful at every step. Their uncompromising quality standards have made this book an exceptional effort. Their encouragement from time to time has been an inspiration for everyone.

The publisher and the editorial board hope that this book will prove to be a valuable piece of knowledge for students, practitioners and scholars across the globe.

Index